# IPAD
# FOR SENIORS

*The Most Complete Easy-to-Follow Guide to Master Your New iPad. Unlock All Its Features with Step-by-Step Illustrated Instructions and Useful Tips and Tricks*

**Gary Watts**

# GET YOUR FREE BONUS NOW!

To thank you for buying this book and to motivate you to study all the features and functions of your new iPad, I am also happy to gift you the digital version of my last guide for the Apple Watch. I'm sure it will help you reach the level of knowledge of modern technology you desire. Enjoy!

**APPLE WATCH FOR SENIORS**

The Most Intuitive Guide to Master Your New Apple Watch from Scratch. A Detailed Manual with Step-by-Step Instructions and Useful Tips & Tricks

## TO DOWNLOAD YOUR BONUSES SCAN THE BELOW QR CODE OR GO TO

https://garywatts.me/bonus-i-pad/

# Table of Contents

Introduction ........................................................................................................ 8

*Chapter 1:* Terminology ..................................................................................... 10

*Chapter 2:* Setting up ipad ............................................................................... 13

The Basic ...................................................................................................... 25

The Lock Screen .......................................................................................... 25

The Home Screen ........................................................................................ 25

Gestures and buttons .................................................................................. 25

How to use Control Centre ......................................................................... 26

Search your iPad.......................................................................................... 26

Searching Your iPad: What You Need To Know. ................................. 26

Searching Your iPad: How Things Work. ............................................. 27

Type with the keyboard ............................................................................. 28

How to copy, paste, and select text .......................................................... 29

Widgets ........................................................................................................ 29

Use Focus mode to prevent distractions ................................................. 39

Passwords .................................................................................................... 39

How to use the Share sheet ....................................................................... 39

Use AirDrop to share files ......................................................................... 41

Control an Apple TV ................................................................................... 41

Siri ................................................................................................................ 42

How to multitask ........................................................................................ 43

*Chapter 3:* Ipad app store & apps .................................................................... 44

APP STORE ................................................................................................. 44

How to Create a Gmail Account ............................................................... 45

How to create an AppleID .......................................................................... 46

Installing Apps From the App Store ......................................................... 46

How to Download Apps ................................................................. 47

How to Remove Apps ................................................................. 47

Popular Apps ................................................................. 48

Utility Apps ................................................................. 49

Productivity Apps ................................................................. 49

Social Networking Apps ................................................................. 50

Popular Games for iPad ................................................................. 51

Shopping Apps ................................................................. 51

*Chapter 4:* Web and communication ................................................................. 52

How to connect to the Internet ................................................................. 52

Browse The Internet With Safari ................................................................. 52

Bookmarks ................................................................. 56

Download and manage files ................................................................. 56

Send an email ................................................................. 56

Chat using Messages ................................................................. 56

Manage your calendar ................................................................. 57

How to Use FaceTime on iPad ................................................................. 58

Imessage ................................................................. 64

Send or Receive Money ................................................................. 65

Whatsapp ................................................................. 65

Creating Accounts in Social Media Networks ................................................................. 66

*Chapter 5:* Maps, notes and utilities ................................................................. 68

MAPS ................................................................. 68

Notes ................................................................. 69

Create Reminders ................................................................. 73

Select & Translate Live Text ................................................................. 73

*Chapter 6:* Camera and photos ................................................................. 75

Sir and Apple Music ................................................................. 81

How to listen to Music ................................................................. 81

Apple Music Settings .......................................................................................82

Watch TV and Movies .....................................................................................83

Podcast on your iPad ......................................................................................84

*Chapter 7:* Ipad accessories ..........................................................................85

Airpods and Earpods .......................................................................................85

Apple Pencil .....................................................................................................88

How to use Scribble .........................................................................................90

Quick Note .......................................................................................................91

*Chapter 8:* Ipad security ...............................................................................92

**Two-Factor Authentication** ......................................................................92

**Turn on two-factor authentication** ..........................................................93

**Turn off two-factor authentication** .........................................................95

*Chapter 9:* Setting and troubleshooting .......................................................97

**Secure your passwords with iCloud Keychain** ........................................97

**What is iCloud Keychain?** ....................................................................97

**Enable iCloud Keychain** ......................................................................97

**How to access passwords using iCloud Keychain** ...............................98

**How to delete passwords from iCloud Keychain** ................................99

**Screen Time** .............................................................................................99

**The Screen Time Dashboard** ...............................................................99

**App Limits** ..........................................................................................102

**Always Allowed** ..................................................................................104

**Content & Privacy Restrictions** .........................................................105

**Other features** .....................................................................................106

**How to recover or reset your Passcode** .................................................107

**Resetting your iPad to factory settings** .................................................108

*Chapter 10:* Tips and tricks for your ipad ...................................................110

*Chapter 11:* Faq ...........................................................................................113

Conclusions ...................................................................................................116

# Introduction

Apple iPad, a tablet from Apple Inc. equipped with several input devices to help you interact with it, is also known as an 'iPad'. This Apple iPad user guide will give you information about the different parts of the device, how each one functions and, more importantly, how to use it.

It has a touch screen display and two cameras; one in the front and one on the back. The front camera has a resolution of 1.2 MP. It also has Wi-Fi connectivity, Bluetooth connectivity and 3G or 4G LTE cellular data capability, allowing you to connect to your home network or mobile internet provider via wireless broadband.

This device is powered by a 64-bit 1.4 GHz dual-core processor, 1 GB of RAM and 8 or 32 GB of internal flash memory. It also has an A4X chipset for processing graphics and video. Though it does not have a rear camera, it has two cameras on the front and a FaceTime HD camera that allows you to use video calls with friends and family over Wi-Fi or through cellular data.

This Apple iPad user guide will help you understand the different parts of this device, how they work and, more importantly, how to use them. You will also learn about all the applications on the Apple iPad and how to install them on your tablet using iTunes technologies. We will also teach you how to activate and deactivate certain applications using the iPad's onscreen menu.

This guide has been created in a very simple way; thus, it is not only easy to use but also easy to understand. We hope this guide helps you learn about all the different parts of the Apple iPad and how they work to get the most out of your tablet.

Many external accessories can be used with your Apple iPad to enhance its features. We will discuss them in detail later.

We will now talk about some of the hardware parts of the device. We will begin with an introduction to the device's display and how to use it.

The Apple iPad has a 9.7-inch touch-sensitive IPS LCD capacitive touchscreen that displays 16 million colors at 1024 x 768 at 132 ppi. It means that it is quite efficient at displaying sharp text, images and videos, but this depends on the quality of your network or internet provider and whether you are using Wi-Fi or cellular data.

If you are using cellular data, it will also depend on the speed of your connection. This display has a wide screen of approximately 4:3 ratio is an ideal medium for surfing the web, playing games and viewing movies.

The Apple iPad features a front camera that also acts as an iSight camera with a 1.2 MP resolution, 720p FaceTime HD video-calling capability and geo-tagging, and dual microphones for surrounding sound during video calls. This camera does not have flash capabilities, but you can use the rear camera for taking pictures or recording HD videos.

The Apple iPad features a rear camera with a resolution of 0.9 MP but a tap-to-focus feature. This camera can take pictures and record videos for up to 1080p data at 30 frames per second with stereo audio and 720p HD videos at 30 frames per second.

The icon on the top left of the home screen from which you can access all your installed applications is called 'iPad'. This represents the device itself and is a very simple way to launch an application without digging through the onscreen menu. You can also launch applications from anywhere though this will depend on where you are in terms of installed apps.

# Chapter 1:

# Terminology

- Widgets: Small programs that can be added as widgets or removed from an existing application (used as additional features). Examples of widgets include the calculator widget, stopwatch widget, weather widget and many more.
- iBook: "iBook" is the term Apple uses to refer to the iPad.
- Air: Air is a given name for the Wi-Fi networks that iPads can connect to. You see it anytime you connect your iPad to a wireless network.
- Airborne: To be airborne means to leave the ground somehow and is often used when referring to flying in space or an airplane.
- APN: The Access Point Name setting of an iPad refers to all the names used for networks that a particular iPad connects and shares with when internet sharing is enabled in Settings > Wi-Fi & Network > Airplane Mode.
- Springboard: The springboard is the app launcher at the bottom of an i-Pad. It allows you to switch between applications quickly. To access the springboard, press and hold the Sleep/Wake button at the top left of your iPhone and use a swipe to slide down to "Home".
- Music Library: This is a term used to refer to all of your music in iTunes on an i-Pad.
- Control Center: Control Center has several modes that allow you to quickly control certain aspects of your i-Pad, such as adjusting the volume or turning it on or off. You can wake your device from Control Center using Touch ID or Siri, saying, "Wake Up."
- Home Screen: The home screen is a page on which all your apps are located. It can be accessed using the springboard.
- Dock: Apple calls the dock a place where you keep your most-used apps. You can access it anywhere using the springboard.
- Album: This is a term used to refer to one part of an album. For example, if you have an album called "Yoshi's Story", each song would be a different "album" by default.
- Now Playing: When you access your music library, Apple lets you play songs in your music library right there and then without having to enter anything else.
- Genre: This refers to all artists under a particular genre. For instance, all artists in the "r-b" genre would be considered part of that genre.
- Track: A track is also a song or music.
- Siri: Siri is an intelligent assistant that can be used on an i-Pad among other things. To use it, simply hold down the home button on your i-Pad or Mac and say something out loud. Siri

will react and support answering straight from the i-Pad. Some examples of what you can do are: "What is the weather like today?" and "Who is the president right now?"

- 2x: 2x means doubling, so a 2x increase in volume means that your volume would be doubled. For instance, if your iPod volume is set at 10% (which it probably is), changing it to double becomes 20%.
- 3D Touch: 3D Touch refers to Apple's new technology. Using your finger, you can press hard on iPhone displays to access apps or get more information about items displayed on them.
- AirDrop: AirDrop is a feature that allows you to receive files over WiFi or even share them via Bluetooth to another iPhone with AirDrop enabled.
- One Pass: In Ad Hoc mode, you can transfer only one file simultaneously. For example, if you want to send a picture in One Pass, you'll be limited to sending one picture at a time.
- Ad Hoc: Ad Hoc is Apple's wireless networking protocol that allows multiple devices to connect and share data without a wireless access point. This can be used by multiple devices such as an i-Pad or iPhone and an iMac or another Mac, which are connected wirelessly.
- AirPlay: AirPlay is a feature that allows you to share content with Apple TV, iTunes and much more. This feature will allow you to stream content wirelessly to another device.
- SIM Card: SIM cards are small cards inside many devices such as cellphones. They allow your device to access the internet or connect with your carrier's network.
- Antenna Band: The metal strip that wraps around the back of an i-Pad is known as the antenna band. It allows Wi-Fi signals to pass through and send signals from one end of your i-Pad to the other, which is usually near the top where most antennas are located, so they do not interfere with each other.
- AT&T: AT&T is a company that provides wireless communication services.
- Bluetooth: Bluetooth is a technology that allows devices to connect wirelessly. It's used by many devices such as phones, i-Pads and tablets, computers, printers and more.
- WLAN: WLAN stands for Wireless Local Area Network. It's what your WiFi connection provides you with (provided you have internet).
- DMZ: DMZ stands for Demilitarized Zone. This is a place on your WiFi network where you can forward incoming requests if, for some reason, your router won't accept them directly.
- WWAN: WWAN stands for Wireless Wide Area Network. It's an alternative way to connect to a network, especially in areas with no WLAN or DSL.
- Safari: This is the default browser on i-Pads and iPhones. It's what you access when you open the Safari app from the home screen.
- App Store: The App Store is always popping up in your newsfeed on Facebook and it will give you access to your own personal App Store, where you can download iOS apps that work with your device, including apps that are integrated with other services such as Gmail, Google Maps, Facebook and more.

- Made For iPhone: This is a term Apple uses to make devices that can only be used on iOS devices, such as iPhones and iPads. In some cases, this means the device will not work with other operating systems and when you purchase it, you'll see a disclaimer saying as much, or you may have to pay more.
- Manufacturer warranty: This is part of the terms and conditions for buying an item such as an i-Pad or iPad Mini. It states that if something goes wrong within the first year of ownership, Apple will fix it for you at no charge.
- 3G: 3G is the third generation mobile communication technology for wireless data transmission. It allows data transfer rates of about 384 kbps, which is not slow but not the fastest either. Many upcoming devices are expected to have even faster 4G wireless internet, including i-Pads and iPhones.
- First name: This is the first name used in your address book or contact list on i-Pads or iPhones. The first name field can also be changed using Control Center and Siri.
- Second name: Your second or middle name can be added here as well as Control Center, Siri and even Contact photos in your address book (if you previously set them up).
- Address book: The address book is where you can store your contacts. You can manually add them to i-Pads and iPhones, by email or search for them on the internet.

# Chapter 2:

# Setting up ipad

When you turn on your iPad for the first time, you will be asked to select some settings like selected language, Wi-Fi network, Apple ID and more. You don't have to answer all these many questions when you are setting up. You can skip some of them now and change them later.

After opening the iPad Air 5 from its box, click and hold the power button to turn it on. Wait for the operating system (OS) to finish booting, and then follow these steps:

1. Please select the language of the operating system. Most third-party applications you install will also reflect this language setting.

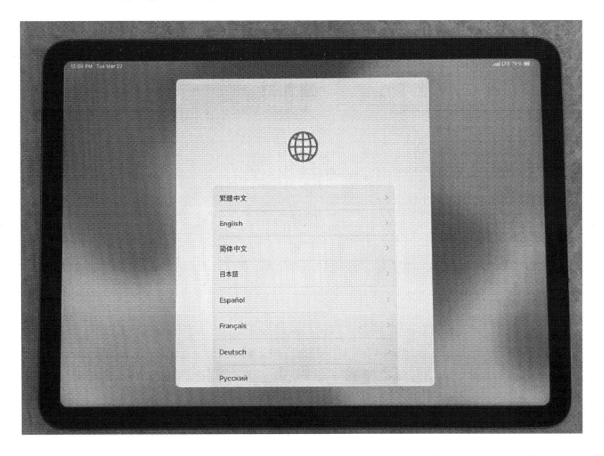

2. Please select your region or country. This will adjust the service and features accordingly.

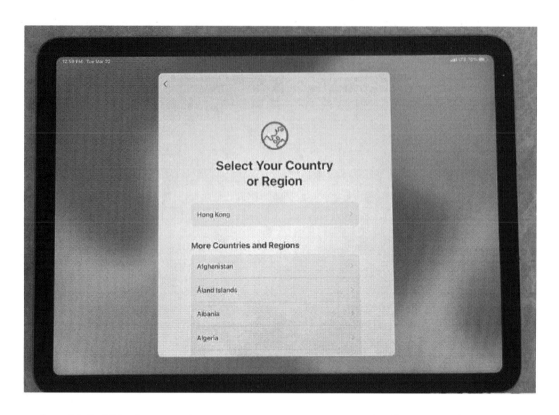

3. If you have another iOS / iPad operating system, you can move it closer to the new iPad Air 5 to transfer data quickly. Otherwise, click Manually Adjust to start from scratch.

4. Please check your language setting. You can make all of the necessary changes by clicking on Customize settings.

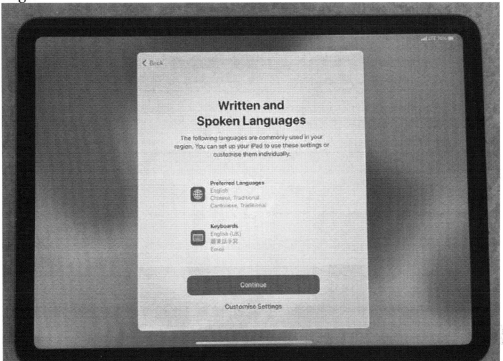

5. The Preferred Language setting allows you to select the priority of the spoken language. For example, if your app does not support English, it is by default a traditional Chinese language.

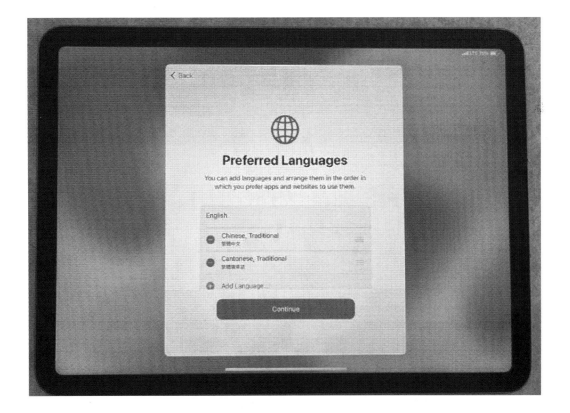

6. Select the language you want to add to your iPad keyboard and click Continue.

7. Dictating allows you to turn your speech into text. Select the language you use in Dictation and click Continue.

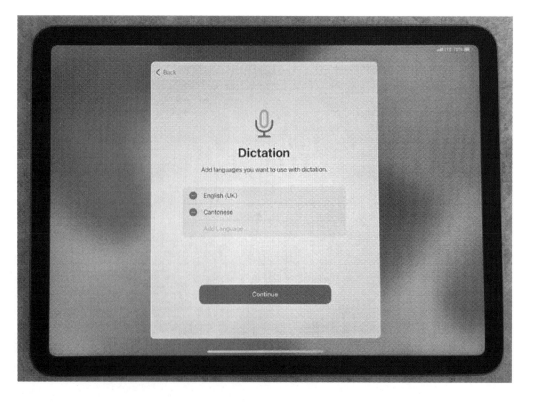

8. Click on your name and enter your password to connect to your Wi-Fi home network. To finish the installation procedure and utilize the functionality offered by the iPad OS, an Internet connection is necessary.

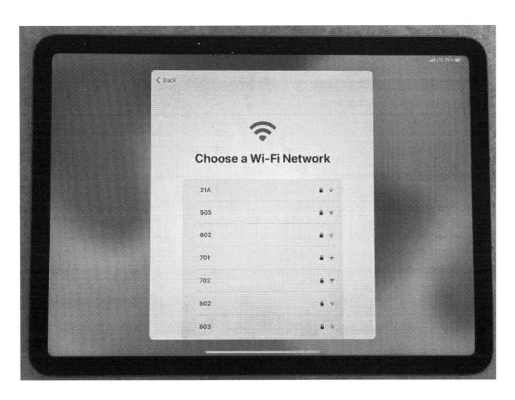

9. It can take up to a few minutes for your iPad to connect and activate your Apple server. This is to ensure you have not been robbed and start counting down the warranty that comes with it.

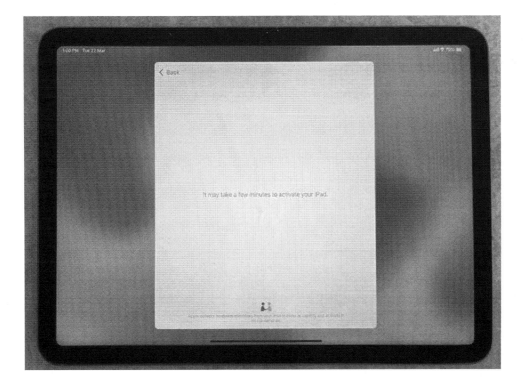

10. Carefully read Apple's data and privacy policy and click Continue.

11. Set Touch ID as an option. This allows you to unlock the iPad Air 5 by placing your finger on the power button.

12. To set the Touch ID, release the switch repeatedly with your finger to record your fingerprint.

13. After setting up Touch ID, you must select a password as a security measure. Using the Password Options button, you can change your passcode from a numeric passcode to an alphanumeric password. Once you have determined the type of password, enter it twice to confirm it.

14. Decide whether to restore your cloud or local backup from another device. Alternatively, click Do not transfer applications and data to start from scratch.

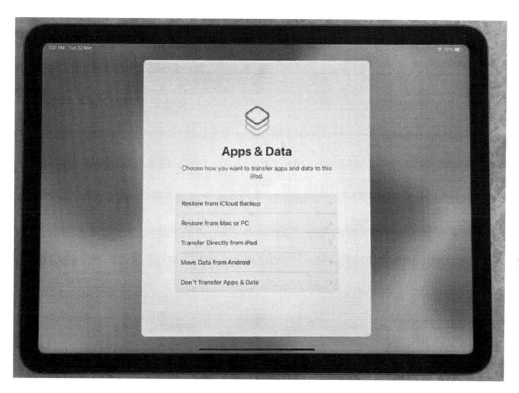

15. Sign in to your Apple ID to access company features like iMessage, FaceTime and App Store downloads.

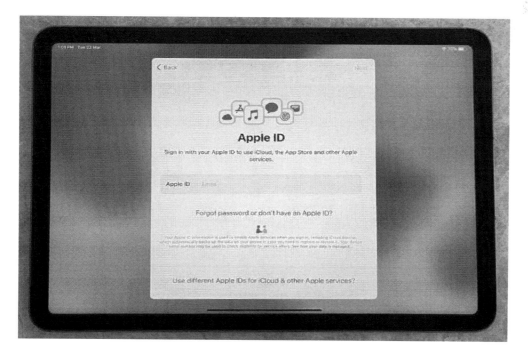

16. Please read the Terms of Use carefully and click [Agree] if you agree. If you do not agree, you will not be able to use the new iPad Air 5.

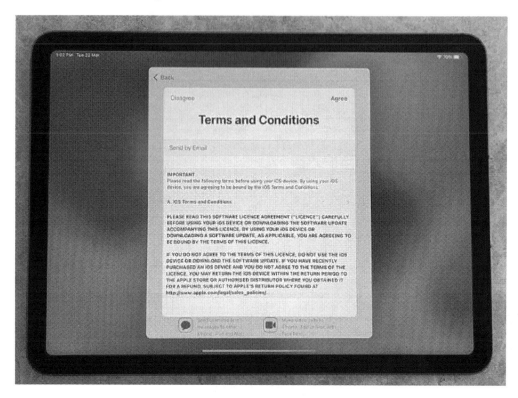

17. Click Continue to get future automatic iPad OS software updates. You can disable this option later in normal settings.

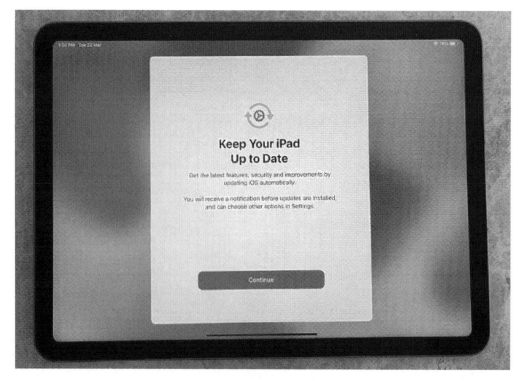

18. If necessary, enable location services to help your app provide more accurate information and use navigation services and maps.

19. If you like, you can enable Siri, a virtual assistant. Siri can answer your questions, perform tasks and provide relevant information on time.

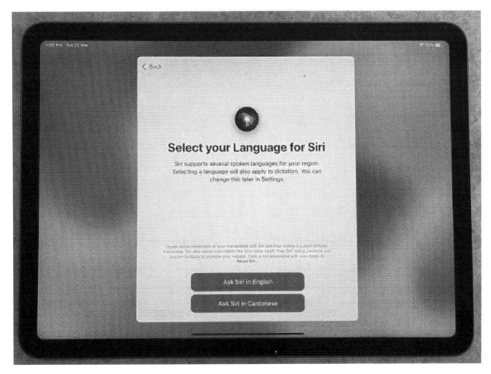

20. If you enable Siri, decide which language you want Siri to use.

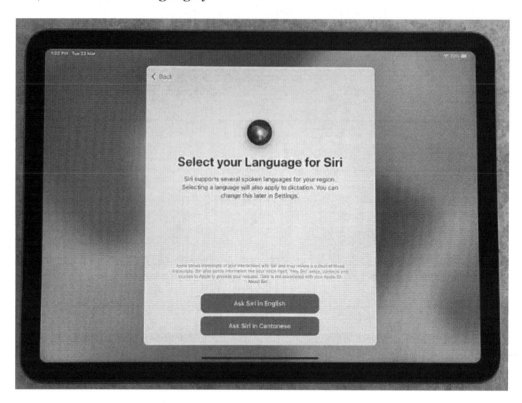

21. Select the sound you want to use with Siri. The included voice has different emphases and human characteristics, so you can choose the voice that suits you.

22. Activate HeySiri if necessary. This allows you to call Siri hands-free by saying, Hey Siri.

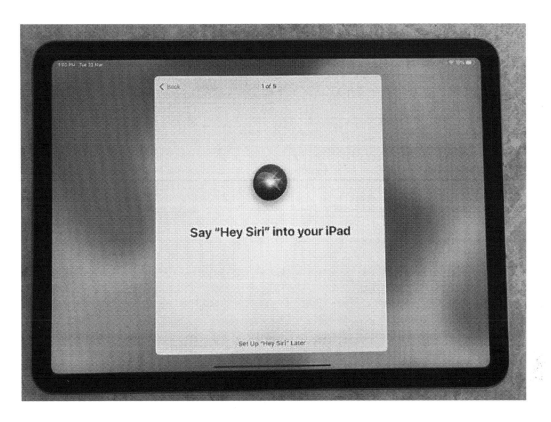

23. Choose whether Apple should collect audio recordings to improve Siri. If you decide to share them, your data will be anonymous and will not be linked to your credentials.

24. Screen time monitors the time spent on iPad Air 5 and sets optional limits and child supervision. You can optionally click Continue to enable this feature.

25. Apple collects anonymous logs from your iPad, which it uses to improve its products and services. You can choose to allow or block this.

26. Decide if you want to use your iPad in light or dark mode. You can later select the automatic option in the display settings. The Dark mode adds a gray/black background to the system and supported apps and the light mode makes it a lighter color like white.

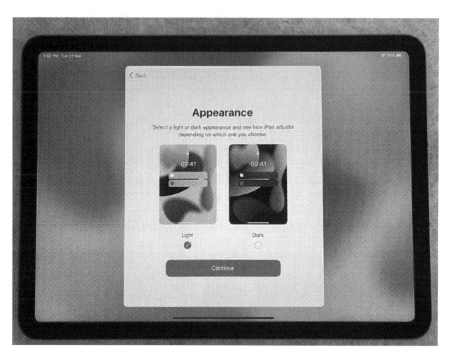

27. Click Getting Started to get started with the iPad.

28. Congratulations!

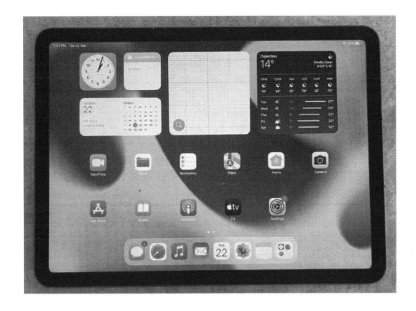

# The Basic

## The Lock Screen

The Lock Screen on the IPAD is a security measure that locks your device to keep others from accessing your information. The Lock Screen will activate if the device has not been touched or unlocked in a predetermined time. As soon as the device is unlocked, you will see an image that looks like the primary screen on your device.

NOTE: The Lock Screen does not protect against lost or stolen devices. You will need to set up a lock screen passcode on your device to protect sensitive data from unauthorized users.

## The Home Screen

When you initially turn on your iPad, the home screen is what you see. It displays all of your apps, a search bar and access to settings. Each app on the screen has a wavy icon, which means it can be opened by tapping that icon; this will take you to the opening page for that app, where you can see what other pages are available. On the screen's top is where you'll find the search bar and it allows you to find any of your apps by typing in some keywords from their name or general category. If you're trying to find an app, not on the screen (perhaps you don't remember what page it's on), you can use the search bar to find it. The settings button allows you to change some basic properties of your iPad, such as turning on/off a Bluetooth connection or opening up a wireless network.

## Gestures and buttons

Since the iPad was released, Apple has continuously updated the device with new gestures and buttons. Even once you know them all, you will still find yourself doing things inadvertently, leading to frustration. This guide will cover all of the most common gestures and buttons on a new iPad mini 2 that are pre-installed by default. It's crucial to note that I frequently use these gestures and button combinations on my iPad Mini 2. Many gestures and buttons come installed by default on the iPad, but I have not covered them as I am only cover those I regularly use.

My favourite gesture is tapping on the screen at an angle and then releasing your finger. This motion allows you to slide between apps without repeatedly using the multitasking button in between apps. As this gesture, it doesn't work with every app, but it allows you to navigate between apps quickly and efficiently.

The swipe-up will bring up the control center when you are in any app. No matter if the display is locked or not, it functions. The control center allows you to perform certain functions like turning on Wi-Fi, Bluetooth, and turning on and off the display's auto-lock feature. The bottom of the control center also contains music controls if you have your iPad set to play music with your media player (Spotify, iTunes) or if you have a song playing in the background.

It is important to remember that you can access additional buttons and choices in the control center for some programs.

## How to use Control Centre

With the release of IOS -11, a new feature called Control Centre was added. It allows you to quickly access important functions such as the Airplane Mode and Wi-Fi, brightness and volume controls, flashlight, calculator, camera and more by swiping up from the bottom edge of your screen. It's incredibly convenient to manage your device without going back through apps or settings menus each time.

Apple seems to be pushing forth the idea that it is working hard for its customers in order for them to have a seamless experience when using their devices. For some users, it does seem like everything has been thought about with how Control Centre works so seamlessly with other features on Apple devices.

But for some people, it can be more confusing than helpful.

The fact that it is implemented on the lock screen means that you could accidentally wake your device up when checking your phone if you are not careful. When you consider the surge in popularity of Apple Pay, this is particularly accurate and contactless payments since the release of IOS-11. Control Centre can be accessed with a single tap by holding down the side button on your device.

## Search your iPad

Searching Your iPad: What You Need To Know.

Accessing The Tool Bar.

To search your iPad:

1. Tap the Home button. This is at the bottom centre of your iPad's screen.

2. At the bottom of the screen, you'll see a bar with four icons. The leftmost of these is Search, which looks like an eye inside a magnifying glass. Tap that icon to get started searching for something on your iPad.

3. Tap the Settings app. This is found on the iPad's main screen and looks like a bunch of squares.

4. You'll see an icon at the top corner of this screen that looks like gears rotating and a horizontal line called sliders. Tap these buttons to access different settings on your iPad.

5. Tap Notifications in Settings to manage notifications on your iPad, which means when you get a new mail message, a new text message or an incoming phone call — or even just a notification that you have new email messages.

6. Tap the General panel in Settings to adjust settings for the iPad's Wi-Fi and other settings.

Searching Your iPad: How Things Work.
Your iPad's main screen, which is called the Home screen, has these features all arranged on it:

The four buttons on the bottom of your screen have different functions than those on your iPhone or iPod touch. The Search button has a magnifying glass, so you can search through documents and apps instead of just using an app, like Safari or Mail, that already keeps all your documents and apps up to date. The three buttons at the top of the screen are Home, Lock and multitasking.

Viewing Items On The Tool Bar.

To access the app or document you want to view, tap it. To sort by date or other criteria, such as category, tap the buttons on the toolbar at the bottom of your screen and select a sorting option there. You can also change how apps appear on your toolbar by accessing Settings, tapping General simultaneously, and selecting a different order for apps.

Viewing The Contents Of Your iPad's Document Viewer.

The documents you create in iWork for iPad apps will be saved to your Documents folder in iCloud Drive and stored until you delete them from your iPad. You can view these documents by tapping the Documents icon on your iPad's toolbar.

Sorting Your iPad's Tool Bar By Date.

This is the how you can organize your iPad to make it easy to find what you're looking for without having to do a lot of searching. It generally works best with apps, like Mail or Safari, which have categories of some sort, so it's easier to find something you need than if you don't have categories on your toolbar. Categories usually are on the left side of your screen just above X-Ray and right below Photos.

## Type with the keyboard

The type keyboard is the most powerful iPad feature- it allows you to type what you think. This guide is designed to show you excellent tips, tricks and shortcuts for typing with your iPad. Your fingers will dance across the keys with ease! After you have mastered the keyboard basics, try some shortcuts to boost your productivity.

You can see the keyboard on your iPad- at any time by sliding it up from the bottom of the screen. The keyboard will appear like a dock with a small light where you enter text. Your iPad will turn on the light when you need to type text. If you have a flip cover, you will have to either take out your iPad or pull it down on one corner of the cover in order to access it.

1) We'll start by getting an overview of where all the keys are on the keyboard:　←↓→↑↓←↑ "A" "S" "D" "F" "G". You should have no problem finding these letters on a QWERTY keyboard.

2) Next, we'll look at how keys are typed: Double tap the key you want to type. The iPad will then automatically type the letter on the keyboard. If you want to capitalize [press Shift Key first], double tap the key twice.

3) You can also type symbols, numbers and special characters with this handy feature: Once you've found where a symbol is located, simply press and hold on to it until a menu appears, then choose a character option. For example, if you want to type '@', press and hold down on [the letter 'A']. Then pick '@' from the menu that appears.

4) The keyboard has special keys to help you type the most useful symbols and characters. To find out more, tap the [?] key on your keyboard. You will see an explanation of each symbol. To learn more about the function of each key on the keyboard, tap on it until its definition appears in a small box at the top of your screen.

Gives you access to tons of other functions

5) You can also get help typing commands in numerous apps: Just tap on '?' to get a list of available qualifiers for apps like Pages, Microsoft Word and others. Tap on the desired app and choose the command you need to use in that app.

6) Tap the '?' key again to get a list of commonly used shortcuts for apps like Safari, App Store, Camera and more. You can also pick the one you want to use for customizing your particular shortcuts.

7) Finally, if you want a list of all keys, why not tap on '?' again? The iPad will automatically highlight each key as it is introduced to make it easier to access.

## How to copy, paste, and select text

Selecting text on the iPad is like selecting text on your computer. The first step to selecting text on the iPad is holding it down and dragging your finger down or up. If you want to copy, release the key or button you're holding down on.

To paste text, press and hold down on a word in an app. You'll see some options that appear at the bottom of your screen. Tap Edit and Paste. The cursor will move to and stay in the location where you tapped.

To select multiple words, hold down on a word and drag it horizontally across multiple words. You can also hold down a word and drag it vertically or diagonally. You will see a blue highlight appear on the words you're selecting. When you release your finger, you'll see some options at the bottom of your screen. Tap Select, Cut, Copy, or Paste.

Pinch to Zoom

To zoom in on something in a document, email, or web page, put two fingers together on the glass and spread them apart. To zoom out, pinch with two fingers and move them closer together.

Use keyboard shortcuts to enter symbols

Hold down on the key to the left of the spacebar until you see the list of symbols pop up. Then use your finger to tap one of them while still holding down. Here are a few handy shortcuts:

Shift + C: Capitalize the next letter

Ctrl + C: Capitalize the previous letter

Alt + D: Delete a letter

Shift + B: Backspace

Alt + B: Boldface current text

Ctrl + [ ] : Select all text between these brackets (optional)

## Widgets

Widgets have become a huge tool for iPad users thanks to the latest iPadOS 15 improvements. Your Home screen and Today View can be customized with widgets so that you can see important notifications and data at a glance. We'll go through widgets in detail on the iPad in this section.

### Add Widget to Home Screen

Previously, you'd only add widgets to the Today View on your iPad. You now have another option: the ability to add widgets to your iPad's Home Screen.

- Start by long-pressing on an empty spot on the iPad's Home Screen page where you wish to add a widget.

- Go ahead and click on the + button in the lower-left corner.

- Follow up by selecting the application for which you wish to add a widget, for instance, the Weather app or the Reminder app.

- Move on by swiping left and right across the various sizes and categories of widgets until you arrive at the style you prefer to use.

- When you find the right one, click on **Add Widget**.

- That's it. The widget will now display on your iPad's Home Screen. You can then go ahead and reorder widgets and drag them across the screens.

- You can move a widget by pressing and holding down on a blank spot on the iPad Home Screen.

- Then, press and move the widget to your preferred position, and let go of it to save its location.

- After organizing the iPad to your taste, click "**Done**" at the top-right corner of the display.

## Add Widgets in Today View

Widgets can be added to your iPad's screen to make it more organized, colorful, and easy to use. To add a widget to your iPad, follow these guidelines:

- Start swiping right until you get to the last screen on your iPad Air. This screen is called "Today View."

- Press and hold down an empty section of the display.

- Then, click on the gray + button in the top-left corner.

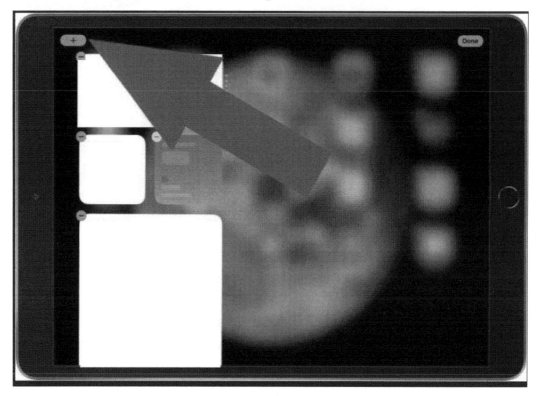

- Follow up by choosing or searching for a widget.

- Go ahead and swipe left or right to select a preferred widget size.

- Then, click **Add Widget**.

- Once done, the newly added widget will appear in your Today View.

- Next, click **Done**.

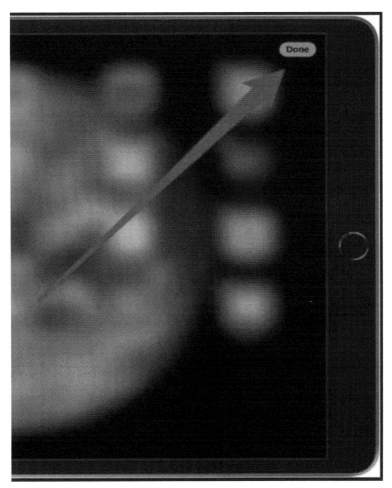

**Remove Widgets from Today's View**

If you wish to get rid of any widget from the Today View, here are the procedures:

- Start by long-pressing on a widget to access the quick actions screen.

- Then, click **Remove Widget**.

- Next, click **Remove** to approve the widget deletion.

**How to Stack Widgets**

Widgets come in several sizes. However, you can stack widgets that are of the same size. The benefits of stacking widgets are that it occupies fewer spaces and you can move across the widgets anytime you want. Follow the procedures below to stack widgets on your iPad:

- Press and hold down on the widget you wish to drag into a stack.

- Follow up by dragging the widget and dropping it at the top of a different widget of equal size to generate a stack or include it in an already available stack.

- Go ahead and add additional widgets by following the steps above. However, you can only stack a maximum of ten widgets. In addition, you'll be unable to combine two stacks. Rather you'll have to move individual widgets into an already available stack.

**How to Edit a Widget Stack**

You can edit a widget stack to customize it to display what you want to appear in it. You can also rearrange and remove iPad widgets from the stack.

- Press and hold down on the widget stack.

- Then, tap **Edit Stack**.

- If you wish to adjust the arrangement of the widgets, move it upward or downward on the list.

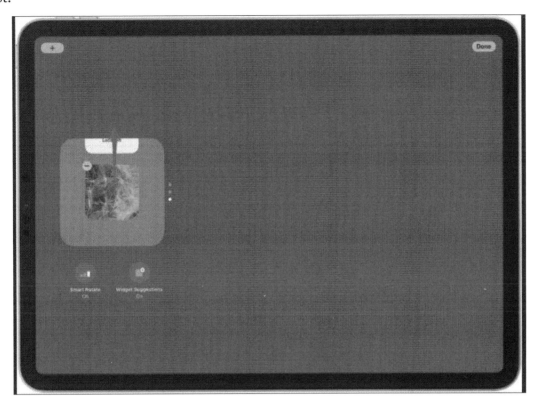

- To get rid of a widget, click the minus button.

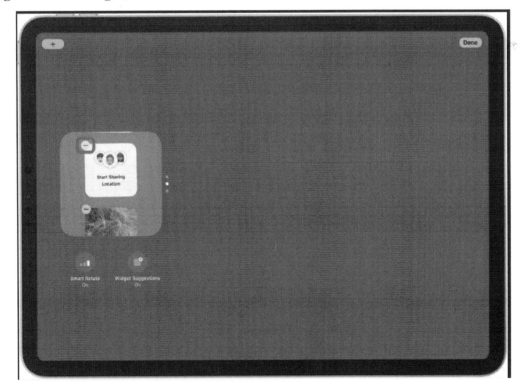

- Click **Remove**.

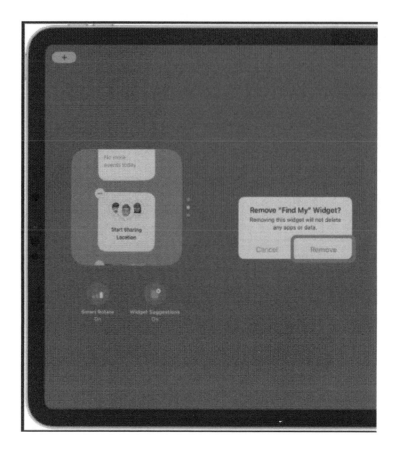

- Then, click **Done**.

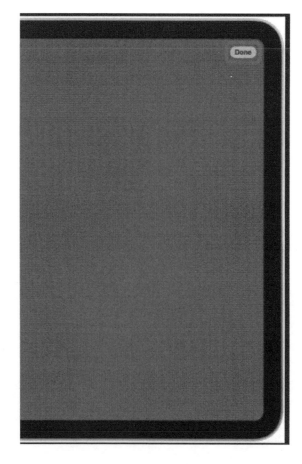

**How to Edit Widgets**

You can customize a widget to display your preferred information. While this is not so for all widgets, the weather widget offers these options. It can show weather conditions in several regions.

- Start by long-pressing on a widget to access the quick actions screen.

- Then, click on **Edit Widget**.

- Follow up by tapping on the settings you want to change and modifying them to your need.

- Once done, click somewhere outside the widget to complete your adjustments.

- After creating a widget in a particular size, you'll be unable to adjust it by editing. Rather, you'll have to remove it and add the newly created one in another size.

# Use Focus mode to prevent distractions

When you use your iPad for activities such as reading or creating content, it's important to set yourself up for success by eliminating unnecessary distractions. The default settings for the iPad allow you to be informed of all incoming messages, but sometimes these notifications will prove too much. For example, a new email alert is the equivalent of an octopus grabbing your attention and not letting go until you've acknowledged its presence. If you're not careful, this can quickly lead to distractions that leave you less productive and more frustrated.

The Focus mode allows users to temporarily disable all incoming notifications and alerts to focus on the task. You'll have to enable the functionality within the General menu of Settings, which will be a toggle switch.

If you plan on using your iPad for work for any extended time, it's highly recommended that you enable this feature to help eliminate as many distractions as possible.

# Passwords

Utilizing a single password across all your internet accounts is terrible, so keep that in mind. Many people use their email address as their password, but you should avoid this because it can be used to hack any contact information associated with that email address. The best passwords are not related to your personal information and are difficult for other people or programs (that might be trying to steal access) to guess. Using words from a dictionary, random letters, numbers, and some personal details ("I love ice cream"), also increases the difficulty of guessing your password.

# How to use the Share sheet

The Share Sheet is a surprisingly powerful tool that can simplify your life in numerous ways. If you've never used it before, here are the basics and how to use it!

Necessary Notes:

- The Share Sheet contains options for sharing content from an app, sending content between apps, and updating information.

- You can also manage what apps have access to share with others by accessing the App Store > scroll down until you see "Share Extensions" > tap the X next to any app you want to revoke permission from sharing.

These are our instructions on how to use this feature on your IPAD.

1. Tap the Share icon in the Control Center

- This takes you to a menu based on what the Share Sheet currently contains. If it's under a specific app, select that one. If it's for all apps, choose "Shared Apps" from the list on the left.

2. Tap an item from this menu to share- A content type will appear in the middle of your screen, along with text and icons for adding or editing that item you chose to share. Your entry can be as simple or as complex as you want (depending on what kind of content you have). Just read through the options before tapping and making your changes!

3. Your new item will be saved when you tap Done on the top of the screen. Any changes made will also be saved after you tap "Done".

4. Share this same item with another app or by saving it to your Photo Album or Email!

5. If you are worried about not saving a certain item, do this:

- Tap on any item in the Share Sheet - The screen drop-down menu should appear and say "Delete Item" at the top of this list (Tap it and confirm). The item will then disappear from the shared section of the Share Sheet. This is how to delete items in general.

- If you want to share something again, tap the back arrow in the top right corner of your screen, and select "Shared Apps".

6. To post a photo or video:

- Go to your camera app and take a shot.

- Then tap the "Post" icon on the Share Sheet (the square with two arrows coming out of it).

- You will open the Camera app and upload your photo there. It will be instantly posted on whatever social site you choose in step #1 (Facebook, Instagram etc.).

- You can also get back to the Share Sheet from your camera app by swiping left or right.

- Make sure you check out our other article on the IPAD - Now what? Everything else you didn't know about it!

## Use AirDrop to share files

Once you have downloaded the AirDrop app from the App Store on your iPad, it will be installed automatically. If you are running iOS 6 or later, you can add AirDrop options to the Control Center using System Preferences under the General tab and enable Turn AirDrop On or Turn AirDrop Off. Select "AirDrop" under General in Menu Bar > Settings to enable it.

Once you are on a Wi-Fi network (or prepared to begin pairing an additional device), having enabled AirDrop on one of your devices, you either shared a file or sent a file using its NFC touch ID or Passcode features (see below).

You will be able to see your iPad and the device that you want to share a file. If you want to send a file from an iPhone, it must be running iOS 7 (or later) and the sender must have an iCloud account. If the receiving device can receive files using AirDrop and is on the same Wi-Fi network as the sender, select it from the list of devices found by AirDrop or by scanning a QR code, then drag and drop the file into its icon or any open folder icon.

If you send files from an iPad to another iPad or iPhone, both devices will need Bluetooth enabled when you first begin sharing. Also, in the options, you can disable "keyboard" sharing so that there will be no need to type a private key for an encryption layer.

A key feature of AirDrop is the opportunity to share with someone else who does not have your device. The other iPad or iPhone must have an iCloud account and be on the same Wi-Fi network as the sender. If both devices are on the same network, they will share files using a QR code scan.

The transfer works even if only one device has AirDrop enabled.

If the other device has AirDrop, you can use two-way AirDrop. If both devices are in the same room, start the AirDrop process by sharing files on one device. When you select the other device from your contacts or your iCloud account, an unsecured Bluetooth connection will be made between your devices and a message will appear on both devices to alert them of the transfer. Afterwards, you can share a file by tapping that icon or share by sending a private message.

## Control an Apple TV

The IPS iPad has been out for a while and is finally available at Apple Stores. It's not as popular as the iPhone, but it still enjoys many uses, including the Apple TV remote control app to turn on your television and change channels or watch iTunes content on your big-screen TV. This requires a couple of accessories. First, you need an Apple TV and an iPad with the Apple TV remote control app installed. Later on in this article, we'll discuss how to install and setup the application.

When you're ready to start using your remote app to control your Apple TV, press the home button on your app - like a mouse click on a computer. A menu will appear at the bottom of the screen, showing all available settings and applications. You can use this menu to search for content, access settings within apps like Settings or General Settings, or access other utilities like Clock, YouTube or Weather.

First, make sure you've connected the iPad to the Apple TV with a cable to control your Apple TV. Make sure that both devices are on the same network. You can check this by going to Settings on the iPad and selecting Wi-Fi. Make sure that you can see your Apple TV listed on this menu. Once everything is connected properly and you're on the same network, launch the Apple TV remote app from your iPad home screen.

Once you launch the application, a menu will appear at the bottom of your screen where you'll see two buttons: one for controlling an Apple TV and another for controlling iTunes content displayed on an HDTV via AirPlay. To control an Apple TV, tap the button to the right of "TV" on the menu and select Apple TV. You can browse through the menus and control most of your installed apps. You'll see various options that allow you to control things like volume, channel selection, play videos or photos and skip tracks (depending on app type).

If you're planning to use your remote to play content from iTunes from your iPad's hard drive, you'll also have access to several options including, Play Music, Playlists or Videos. You can also set these up as new pre-loaded playlists or music for future usage.

Since the iPad is a touchscreen device, you can enter text by tapping individual keys on the virtual keyboard. The virtual remote app also has features that make it easier to use including, "Landscape mode", which makes it easier to touch icons on the screen and "Auto Scroll", which makes text appear faster while you're typing.

In order to turn off your Apple TV, you can simply press the home button on your iPad. This will take you back to your home screen menu and simultaneously turn off your Apple TV.

## Siri

Learn how to ask Siri

1. Create a contact.

2. Press the Add button.

3. According to how you use other apps, such as emails you get in Mail and calendar invitations, Siri will also propose new contacts for you. Show Siri Suggestions for Contacts can be disabled by selecting Settings > Contacts > Siri & Search and selecting that option.

4. Siri also offers contacts in other apps, depending on how you utilize your contacts. See About Siri's suggestions on iPad to learn how to disable this function by going to Settings > Contacts > Siri & Search and unchecking "Learn more about this app."

# How to multitask

Start by looking at what different types of apps do best with multitasking functionality. You'll want to include the most popular apps such as social networking, photo and video galleries, music players, games and whatever else you think your customers will use. If there are any apps that people would regularly use without the ability to multitask, then don't worry because some of these tasks can often be accomplished with a mouse or keyboard at double the speed.

Once you've selected your top apps, you must test each for multitasking capabilities. Look at the app's settings (usually under 'General') and make sure that everything is set up for multitasking before testing it. You'll also probably want to check out things such as gestures and options.

You will want to consider two scenarios when looking at how your apps work with multitasking. One is where you're using a task that's useful in real life and the other is when you're using a task that's not very practical or useful. You'll want to test out both and look for what works best for your selected apps. Take apps such as Pandora Radio which is great for multitasking because it allows users to listen to their favourite music while they do other things on their iPad. Or you could also test out apps such as Facetime which is made for one-on-one conferences, so it doesn't make sense to have your friends waiting for you to focus on the conversation.

The best way to test the iPad's multitasking functionality is by using the app you're testing with a friend. Have your friend use their iPad and give them some tasks while multitasking. You should first see how they handle multitasking when the task is easy and then watch them when it's something that requires attention or a lot of work. Do they get distracted? How quickly can they respond? Regarding testing, we found out that it's best to give your users a task that requires them to think. That way, you're able to see whether or not your user's experience with multitasking is going to be a good one.

# Chapter 3:

# Ipad app store & apps

## APP STORE

In the Application Store, you can find new applications, tips & tricks and in-application activities.

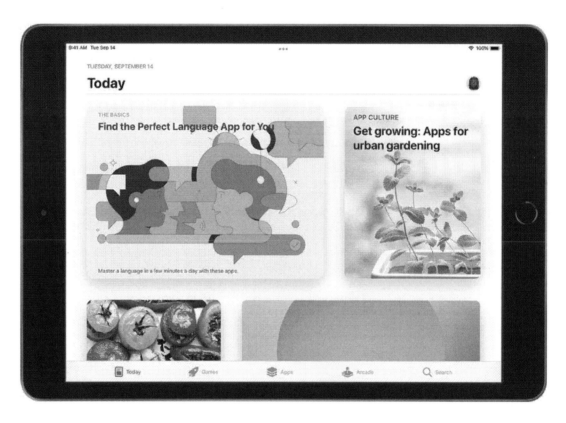

Note: You will need an Internet connection and an Apple ID to use the application store.

### Search for applications

Click on any of the below to find applications, games, and in-application events:

- Today: View applications & in-application events.
- Games: Find games in different categories like action, racing, adventure, puzzles, etc.
- Applications.
- Arcade: Enjoy a collection of premium games from Apple Arcade without ads.
- Search: type what you want, and click the Search button on your keyboard.

### Learn more about an application

Click on an application to view the following info & more:

- Family Sharing Support & Game Center
- Language support
- In-applications events
- Screenshot or overview
- Size of File
- Ratings & comments
- Other Apple devices compatibility
- Privacy info

### Buy & download an application

- Click the price. If the application is free, click the Get button.

If you see the Redownload button ⊕ instead of the price, you have bought the application and can download it for free.

- If necessary, verify your Apple ID with Touch ID or password to complete your purchase.

You can find the application in the Application Library in the recently added category. While downloading the application, an indicator would appear on the application icon.

# How to Create a Gmail Account

Setting up a Gmail account will typically be the first action users take to access their email notifications, which only takes a few minutes. Gmail for iOS can be found on the App Store and is free. There are many options, including the ability to sync your iPhone contacts with Gmail.

Gmail can create email accounts from different places, including Facebook and Google Plus, so you can set up Gmail accounts from all over the place so you won't have to worry about multiple emails. However, if that's right for you, you can also set up your Gmail account when you're just setting up your iPad or even when setting it up if you already have a Gmail account.

Once you have your Google account, you'll be able to create a new Gmail account from your iPad. You can also use the Gmail app if you're already using another email and are just mirroring everything to your iPad or if you want to use the iPad as a device that checks emails while you're out and doing other activities.

# How to create an AppleID

It can be confusing creating your AppleID, but it's never too late to learn. This article is designed for those who have never created an AppleID or those struggling to remember the process. You'll learn how to create an AppleID, complete with step-by-step instructions and helpful tips.

In this informative blog post, we'll teach you how to create an Apple ID and what exactly is required of you when doing so. We'll discuss everything from the basic requirements like creating a new email address and setting up a password, all the way through confirming your information and adding payment information with apple pay. You'll be a pro at creating an apple ID by the time you're done reading this.

The first thing that you need is an email address. If you already have one, then good for you. If not, then all that you have to do is sign up for an email account somewhere and during the process, make sure to select "I would like to receive important communications from Apple."

Next comes password creation. You can follow our suggestions or come up with your variation of letters and numbers, whatever you feel comfortable with, to be honest, because it won't matter in a few minutes anyway. Make sure it's at least eight characters long and contains a mix of letters and numbers.

Now Apple has a limit on the password you can use. It's six characters, with at least one uppercase letter. I've never been able to figure this out but rest assured, it is six characters long and does have an uppercase letter.

After you've set your password, you'll be asked to fill in your birthday and two security questions. The questions are self-explanatory, so we won't go into much detail there, just remember that the answers to these questions have to be something only YOU know. If you put in something like your mother's maiden name, your account goes if somebody gets a hold of the information.

After you've answered your security questions, you'll be taken to the final step. Click on "finish" and when complete, you'll see this screen:

This is where you verify your information and complete the sign-up process.

# Installing Apps From the App Store

Step #1: Load up your web browser and go to https://www.apple.com/app-store/. This will take you straight to Apple's App Store website on your computer or laptop computer (Apple TV users need not worry!). You can also search for the "App store" in Safari from your iPad itself. Enter your Apple ID and password at this point and you'll be taken to a screen that looks like this:

Step #2: Click on the search field at the top right of the page (circled in red above) and enter the name or keywords for what you are looking for in an app. For example, if you want to search for

"Facebook", just type "Facebook" into that search field. You can also browse through the categories left-hand side of this page, but searching is often much easier. Entering something in that search box will bring up several results - click or tap on your desired application (Facebook, Netflix, etc.). The following screen will present a description of the app and any screenshots or user reviews. Click to download the app and you'll then be taken to a screen like this:

Step #3: This is just like the Apple Store for desktop computers, where you'll need to click "Install" (to install the application onto your iPad) and enter in a password if necessary. Once installed, you can go back and either use it on your iPad in the usual way or download it onto any other computer you want to use it on (just make sure that the computer has iTunes or iFunbox installed).

## How to Download Apps

Launch the App Store app on your iPad. Look for the software you wish to download (or search for it). Once found, click "Free" or "Get". A box should pop-up asking if you are sure you want to install it.  If this doesn't happen right away, tap on "Purchases" at the bottom of the screen in order to find them. A pop-up should appear, asking if you are interested in the purchase. Tap on "Yes". The download should start and it may take a few minutes. (Note: In some cases, the app will show up immediately. In that case, just wait a while and it will finish downloading.) Once finished, your app will be downloaded on your iPad.

## How to Remove Apps

The most important thing to remember is that deleting an app from your device does not delete them from iTunes or iCloud. So if you want to get rid of an app on both devices, make sure you use iCloud Drive or find "other ways" in this blog post.

There are two very simple ways to delete apps on your iPad. The first method is called "Method 1", and the second is called "Method 2".

DECREASE THE APP'S STORAGE QUOTA AND IT WILL BE REMOVED FROM YOUR IPAD

To do this, you must go to Settings → General → Usage → Manage Storage. Then click on each app that you want to delete and see if you can see a little white bar in the top right-hand corner of the screen. This indicates that your device has enough room to remove the apps, so long as they are less than 10GB. To manually remove that app, press the " − " button. You need to delete some of your other apps if you cannot remove them.

Once you delete these apps, press the back button on your iPad until you are out of Settings.

METHOD 1: USING APP SWIPE TO DELETE APP

You can delete the app by swiping left on the app's preview and pressing "Delete". WARNING!!! This will result in losing all data for that specific application.

47

You can also delete it by tapping on the app's preview picture's little "x" icon. WARNING!!! This will result in losing all data for that specific application.

Remember that when you delete an app, it will also be removed from your iPad's "iTunes or iCloud" section.

You must use these two methods only to delete apps you have purchased and downloaded using the iTunes store or iCloud. Any other apps, even if installed on your device through a third-party source (which would be an application package installed directly onto your device without help from Apple), would not be easily removed. Only the above methods will allow you to remove these particular applications from both devices. If you want to delete any application from iTunes on your computer, we recommend performing a full backup of your iDevice before proceeding.

## Popular Apps

Here is a quick look at 4 popular apps that may be of use to you:

1) FaceTime - FaceTime is an app for video chatting with other iOS device users. Two-way video chat means both parties can see and communicate online in real-time. This app can be accessed through the camera icon on your home screen and requires an active WiFi connection or data plan to work properly. FaceTime is not free and, as of this, writing has to be purchased through the App Store. The term session describes the time that FaceTime can be used before an Internet connection must be re-established. A 1-hour session costs $0.99 and a 24-hour session costs $4.99.

2) Contacts - Contacts is the iOS app for your iPhone or iPad to track all your contacts, pictures, videos and more. Contacts list all the contacts you have stored on your device, their phone numbers, addresses, email addresses, birthdays, and some fields from social networks such as Twitter and LinkedIn. Using your Apple ID, you can also sync your contacts from Gmail, Yahoo, or iCloud. Contacts can be accessed from the "Pages" icon on the home screen.

3) Safari - Safari is an iOS app that provides a web browser environment for accessing different websites. This app allows you to go to any webpage you choose and contains built-in tabs to have instant access to more than one website at a time. Safari works best with 3G or WiFi service and is designed for iOS devices such as the iPad, iPhone and iPod Touch. You can access Safari from the "Safari" icon on your device's home screen.

4) App Store - The App Store is an iOS app that allows you to buy and download applications for your iPad, iPhone or iPod Touch. You can download apps for free, but you must have a valid iTunes account to make the process easier. This can be done through your Apple ID, or if you do not have an Apple ID, you can create one at www.apple.com/icloud/. To preview and purchase apps for

FREE, you will need to select the "Get Apps" option from the main menu of the App Store app itself rather than buying them directly from within the app itself. Apps are available in various categories, such as books and games.

# Utility Apps

The iPad is a technology that lets you enrich your lifestyle by performing tasks such as browsing the internet, watching videos, playing games, and shopping. Numerous apps allow you to turn your iPad into a personal assistant. For example, if you want to cook dinner but don't know how to cook, the cooking app will walk you through the steps with pictures and easy-to-read instructions.

# Productivity Apps

iWork

'iWork' stands out as the most indispensable iPad app. If you use your iPad primarily for work and do not have 'iWork' installed, there is no point in owning an iPad. 'iWork' will save you time and frustration with many of your daily tasks. For example, many people take notes during meetings by jotting them down in the Notes app on a virtual sticky pad and then syncing it to their iPhone for proofreading. With 'iWork' on your iPad, you can write a note directly onto the virtual sticky pad and do not need to go through the complicated process of exporting your sticky note and importing it (which is why many companies require that all employees have 'iWork' installed).

The key reason you should install 'iWork' is that it saves you time. The apps are extremely well made and only require a few taps to do what they do. It's one app that makes you wonder how you lived without it. The other key reason is that 'iWork' is a platform allowing additional productivity apps to be made for the iPad. If you fit into either of these groups, installing "iWork" is something I'd highly recommend. They can make an entirely new version of 'iWork' by themselves. So everyone wins.

The only reason you probably don't have 'iWork' installed already is that you either (a) don't have an iDevice or (b) are not a person who works with a computer most of the time. If you fit into either of these groups, installing "iWork" is something I'd highly recommend.

For those of you who are worried about jailbreaking your iPad, don't be. The only thing 'iWork' requires you to do is to change the default font, so it won't affect your system files.

iPhoto and Camera

If you are new to the iPad, I'd recommend installing the iPhoto app first (because the camera is built into this app). This allows you to store photos on your iPad and edit them with useful tools. One cool feature of the iPhoto app is that it allows users to create albums of their photos and share them through Twitter or Facebook. I use this app when I plan a party or buy new furniture and want to show my friends.

Another cool feature of the iPhoto app is that it allows users to edit photos made with their iPads. This is useful for editing those 'selfies' you took of yourself from a far-away distance. To be fair, some people might not like this feature as much as others, but it does have its uses. The reason why I bought an iPad was for the camera and iPhoto works very well for me.

Slack

Slack is an app that allows you to send instant messages to your colleagues, making it a very useful tool for the working person. It was also one of the most downloaded apps for iPad in 2014. As the name suggests, it is a messaging application and is, therefore, a useful tool for those who work in groups (if you work with people). Slack is what everyone else has been doing for so long and it can be done on your iPad.

I don't use Slack as much as some of my colleagues because I tend not to have many people I communicate with daily. Since Slack is a group messaging application, it makes sense to me that most workplaces use this app. It's a very useful tool, but it might not be for everyone.

Twitter

The Twitter app is probably the best app out there to tweet and engage in debate with other users. It has the most robust search and discovery functions of any Twitter client on the market and allows users to take advantage of all the features that Twitter has to offer. Even if you dislike Twitter, you should still check out the app as an iPad user. It's a great way to discover other interesting content on your iPad.

Productivity Apps

'Productivity Apps' is another essential app for iPad users. The apps help you manage various tasks that are all useful in their way. To name a few, 1Password (a password manager), OmniFocus (a task manager), FocusWriter (a writing app) and Instapaper (an app that allows users to save articles they are reading). If you install all these apps, you'll be able to store passwords, write emails and read articles without worrying about losing any important information.

# Social Networking Apps

1. Open the Safari app on your iPad and search for "Twitter".

2. Tap on the blue icon and enter your username, password, and email address in the fields that appear. This is known as signing up with Twitter on your iPad or iPhone (complete with attached photo).

3. Press continue to login with Twitter on your iPad, iPhone, or iPod touch.

4. In the top right corner of the screen, you should see your profile photo at the top with your name below it. That means you're all signed up with Twitter on your iPad! You can now begin Tweeting away to your heart's content!

5. Tap the blue icon to see what's happening in the Twitterverse. You'll see a list of tweets (or posts). Below each post is a blue icon with a white bird and what looks like a sideways "V". This is known as an @-reply or a mention. If you want to respond to those people, tap on the @-reply icon.

6. Anything you type in that box is a new Tweet and will be stored in your profile. Tap on the "Tweet" icon on the bottom right of your screen (you can change this if you'd like) and send your first Tweet!

# Popular Games for iPad

They are great for playing on your IPad and are available in the Apple App Store. We will cover Pokemon GO, Crossy Road, World of Goo, GTA: San Andreas, and Angry Birds 2.

# Shopping Apps

*Amazon: Offers everything from books, music and movies to groceries, clothes and jewellery with great daily deals.

*Best Buy: Where technology rules.

*eBay: Browsing, buying and selling at the world's largest online marketplace.

*Kmart: Shop Kmart and discover great values, awesome buys and amazing prices.

*Target: Shop Target to find an amazing selection of products at competitive prices.

*Toys R Us: Be amazed by the toys, games and gadgets kids want to play with today!

* Walmart: Save money. Live better. (Shopping apps have been updated since last year's post).

We recommend you consider several shopping apps and look out for the new iPad.

If you are an iPhone user, you should be able to get some great shopping apps as well. With this QR-code scanner, Google has developed a shopping app that allows you to search the web and look at shelves around you.

# Chapter 4:

# Web and communication

## How to connect to the Internet

First, you need to start downloading the Safari app for your iPad. Once you have, enter the website address, which will turn into a link you can click on with Safari. You're going to need internet access in order to do anything on an iPad.

Connecting to wi-fi is easy and provides a more stable connection than cellular data networks. Enter your password into the 'password' field and accept any warnings that pop up by clicking 'yes'. Once it connects, close out of iCloud settings and then go back to Internet Connection Settings in Settings > Wi-Fi. If a yellow triangle is present, click on it until it says 'i' for information. To properly configure your connection, adhere to the directions provided.

If your wifi connection is still giving you problems, click on 'repair' and follow the rest through. Once finished, try connecting to the internet again.

Assuming you've already downloaded any apps from the App Store (if not, do that), click on one to begin using it. The app will show a message saying 'iPad needs to connect to Wi-Fi'. Tap 'OK' and wait for the iPad to connect. Once connected, open the app you want to use and tap on 'ok' on the lower left-hand side of your screen. It will ask if you want to connect with a shared network or create a new one. Tap 'create a new network' and enter your password. Once you've entered it successfully, your iPad will stop asking to connect again, so work through any prompts.

## Browse The Internet With Safari

### What is Safari?
Safari is a browser built into the iPad. It's similar to other browsers like Google Chrome or Internet Explorer. Use this to access the webpage on your device. Safari's interface is similar to that found in desktop browsers, and it also has some unique features that make mobile browsers easier.

### Get to know Safari

**Navigation buttons**
Use the back and forth buttons to scroll between recently visited pages.

**s / reading lists / shared links**

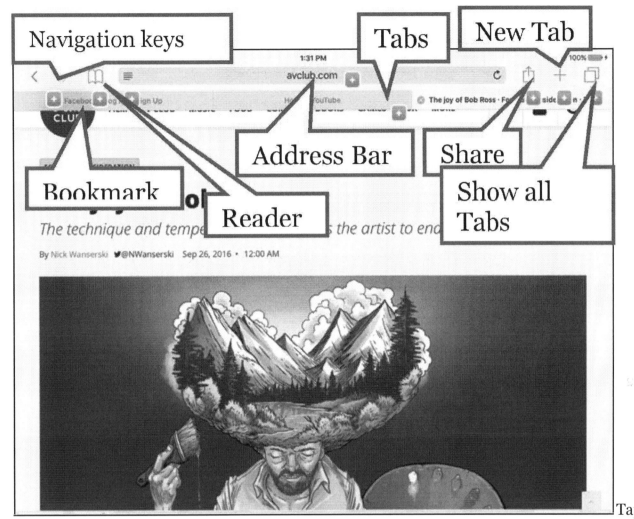

Tap here to view bookmarks, shared web pages and reading lists. You can read web pages that you have added to your reading list even if you are not connected to the internet.

### Reader
The Reader feature lets you view your web articles in a large, easy-to-read format without any ads or clutter.

### Address bar
The address bar appears with the URL of the current page. To go to a new page, just enter the new address. You can also search the web using the toolbar.

### Tabs
Safari simplifies multitasking by allowing you to browse and open links on separate tabs. To go to another tab, tap the tab you want. Tap X to close the tab.

### Share button
You can create bookmarks and save pages to your reading list by clicking the Share button. You can also email the link to your friends or share it on Twitter and Facebook.

### New tab button

Tap here to create a new tab. When you create a new tab, Favorites will appear. This site contains shortcuts to the web pages you visit most often. You can click on the web page to go to that page or enter the URL in the address bar.

### Show all tabs button
Click the Show All Tabs button to see thumbnails of all open tabs.

## Browsing Online
Safari includes several features that make browsing the web easier. We have put together some of the most useful features below.

### Open links in new tabs
Once you have found a link to a website of your choice, you can open it in a new tab. This allows you to browse the linked site without losing the location of the original page.

To open a link in a new tab, press and hold, and select Open in a new tab.

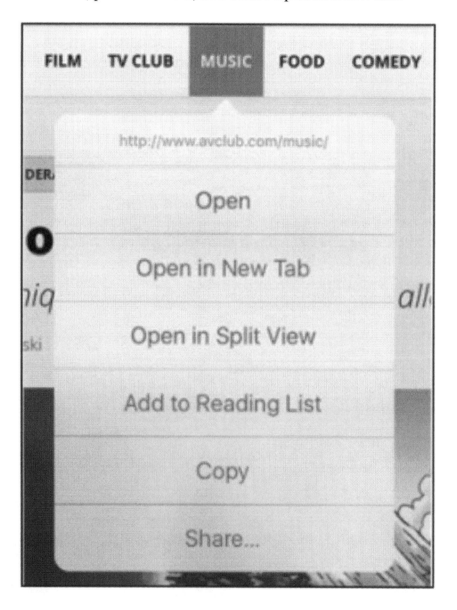

## Multi-touch zoom

It can be difficult to read web pages in Safari. Fortunately, multi-touch gestures have two ways to zoom in and out.

- **Double-tap the screen**: This gesture expands to the area of the screen that is clicked. This is especially useful when reading small text. Double-click again to zoom out.

- **Screen Pinch**: This gesture allows you to zoom in and out to gain more control over the screen size of your page.

## Screen direction

Remember that you can turn on your iPad anytime to change the screen's orientation. While some sites appear better in portrait view, others are better in the landscape.

## Add a web page to your home screen

If you visit the same website frequently, we recommend you bookmark it. You can add icons to the home screen home page if you want faster access. Simply tap the Share button, select Add Home screen and click Add.

This page appears as an icon on the home screen. Tap the icon to open Safari and display the webpage.

# Bookmarks

In Safari, you can save the current page as a bookmark by tapping the "+" button at the top of the browser. A menu will pop up where you can choose to add this bookmark to either your Favourites or Reading List.

You can also manage your bookmarks from a separate list of all your saved bookmarks. Tap the "Bookmarks" icon in Safari to bring up that list.

Here, you can tap on any given bookmark and then tap "Edit" at the bottom of the popup menu to change it, remove it or even mark it as a favourite. Tap "Done" once you're done with your changes.

# Download and manage files

## Send an email

You may use your iPad to send emails in various ways, but this tutorial will focus on using the Mail app. Here's how it works:

1) Open the Mail app and tap New Message in the bottom left corner. Make sure you're connected to WiFi or mobile data and not EDGE/3G for faster service.

2) In Mobile Number, tap next to "+" and type in your contact's number or email address and select Done. Select the appropriate internet provider name in Server, if you have their username. If you don't know what user name to choose, go to your contact's email address and look at the List of servers at the bottom. Look for "google mail" or "gmail" and choose that one.

3) In Subject, type your message and in Message, write the body of your email. There is a paperclip next to the Attachment that you can tap on if you wish to attach a picture or document.

4) Hit Send! You will see a timer next to it indicating how long it will take for your message to send. If your email takes a while to process, check your mail, hoping that it got through.

5) If your message didn't send satisfactorily, try refreshing your email and contact information by tapping the circle icon at the top left of your screen.

If you are writing an email, tap on the pencil icon at the top right corner to access some additional tools.

## Chat using Messages

First, make sure you have the Messages app downloaded from the App Store. If not, head to your home screen and find "App Store" in the list of apps. Click "Messages" in order to download it onto your device.

You will be asked to sign in when you first download the Messages App on your iPad. The app cannot function without an internet connection, so you need to sign in with an Apple ID, the same account everyone has on Macs or iDevices. You can find this through Settings –> iTunes & App Store –> View Apple ID.

You won't have access to all of your messages, contacts, and photos until you have signed in with your Apple ID.

If this is the first time you are using this app, then this may take a while. You will see a dialogue box saying that there are no messages. This is because the app downloads everything, including your contacts and photos.

Once the download has finished, you should be able to see your existing messages on this page. If not, it may be due to syncing with iCloud, which you have set up with your Apple ID.

The top of this screen will tell you that "Messages" is now installed on your iPad. Tap "Messages" at the top of your screen to log in.

This is where you will see all of your messages. If you have used this app on a Mac, you will see your contacts list at the top. If it isn't there, then you must agree to turn on iCloud first. You can see how to do this by going to Settings –> iCloud. Do not enable any other services using this app, or the last few pages may be completely blank!

If your contacts are listed, you can tap on them here and they will appear with their photographs. It is also possible to send them messages from here, but the app is simple and you are given only a few options. It is designed to be used because it has such a large screen, so the buttons on top of their pictures aren't necessary when you can fully use the screen itself!

If your iPad doesn't have a front-facing camera, you will not be able to see photos on other people's devices. If this is the case, they will need to take a picture of themselves and send it to you by email or text message.

# Manage your calendar

Your iPad Calendar app is a powerful tool, allowing you to organize your day, week and month with a single tap. Use it to see who's invited to what meeting, block out time for tasks like grocery shopping or watching TV and even plan the rest of the year.

Here we've rounded up all our best tips for using the Calendar app: * The calendar view has four main areas: Day View, Week View, Month View and Year View. Swiping left on any of these icons will take you back one level. * If you need more space on your calendar at any time, just press and hold anywhere on the calendar until it zooms out, then drag to the space you want. * The iPad calendar app will automatically sync with your iPhone calendar and your computer, allowing you to quickly see what's on your schedule. Sign into iTunes or iCloud to access the full features, including birthdays and invitations. * You can also add or remove events using Siri. Just ask what you want to do by saying "Add event to my calendar" or "Delete from my calendar", for example.

* From calendars that are synced from other devices, tapping and holding an event will allow you to choose which synced calendars it appears in. * Add events straight from the Notification Center by swiping left to bring up the Today View, then tapping on the Calendar button. * To archive an Event, tap and hold until a menu appears. Choose "Archive" from this menu to remove the event from your calendar without deleting it completely. * During meetings and appointments, you can choose to be reminded of upcoming events during that time. Just swipe left to bring up the menu for that event and tap "Remind me at a location", then tap "Remind me when I arrive".

* To manually set the alarm for any event, swipe left on an event in your calendar until a circular 'alarm' icon appears in the corner of that day's square. Tap this icon to access the 'Alarm' screen, where you can choose between a morning, afternoon or evening alarm. * The notification centre can also set an alert for upcoming events. Swipe down from the top of the screen and tap "Notify me for upcoming events."

* In addition to any other calendar, you can add additional calendars to your device, including Google Calendar, Exchange ActiveSync and iCloud calendars. Just tap on Edit in the upper left-hand corner of your calendar and select Add from a computer, iPhone or iPad. * To view more details about an event, such as time and location, use "Details" by tapping on it.

# How to Use FaceTime on iPad

Video calling has been increasingly popular in recent years, and it's a great way to stay in touch with your loved ones even if you can't visit them in person. All Apple devices, including the iPad Air, come pre-loaded with Apple's FaceTime video calling app. With this, your favourite contacts are always within reach, as long as you have Wi-Fi or cellular data to connect with.

## Make a FaceTime Call
- Launch the FaceTime app.
- Now, click on the **New FaceTime** option.
- Follow up by typing in the contact(s) that you wish to call through FaceTime. You can enter contact(s) name by cell phone number, name, or email address.
- To connect to a group through FaceTime, keep entering the contact info until everyone is added.
- Go ahead and click on the Audio or FaceTime Video option to begin your call.

## Turn Off Video or Mute Yourself on FaceTime Call
- Begin by starting or answering a FaceTime call.
- Follow up by clicking on the Microphone or Camera switch in the floating toolbar during the call.

- If you're muted and then you attempt to say something, you'll receive a notification telling you that you're silenced; simply click on it to unmute yourself.

## Create a FaceTime Call Link

You can create a FaceTime link and send it to other people to participate in the call. You can do this for both Apple device users and Windows and Android device owners.

- Launch FaceTime.
- Then, click on **Create Link**.
- Next, click on **Add Name** to give your link a name.
- Input a name for the link.
- Click on **OK**.
- Then, click on the sharing process you prefer to use.
- Follow up by entering the entire sharing info you require for your chosen method.
- Go ahead and send the link through your chosen sharing method.
- The receiver of the link will have to click on the link to participate in the FaceTime call.

## Delete a FaceTime Call Link

You can make a FaceTime call link inactive so that when someone clicks on the link, they won't be able to join the call anymore.

- Launch FaceTime.
- Follow up by swiping left on the link that you wish to get rid of.
- Now, click **Delete**.
- Click on **Delete Link**.
- That's it.

## How to Use SharePlay in FaceTime

Apple has made screen sharing on FaceTime very easy. You can now make other people on a FaceTime call see what's showing on your iPad screens, such as the video you're watching, the songs you're playing, or the website you're currently on. You can even teach them how to fix their devices' settings.

Interestingly, the video and audio are synced across the devices of every participant on the FaceTime call. That removes all of the burdens from the individual who started the video or audio and made it a more pleasant experience for everyone involved.

However, the other participants will have to "Open" a SharePlay by tapping on the option in their FaceTime controls. If they don't join, they'll be unable to view what is showing on your screen.

## Turn On SharePlay

To enable the use of the SharePlay function on your iPad, you'll have to turn on the toggle in the Settings app.

- Launch the Settings app.
- Move down and tap on **FaceTime**.
- Next, click on **SharePlay**.

- Go ahead and click on the SharePlay switch from the new screen and you're set to share videos and audios through a FaceTime call.

## Use SharePlay to Play Music on Facetime

For SharePlay to work on other participants' devices, they'll have to ensure that their device is running on the latest iPadOS, iOS, or macOS version.

- Launch FaceTime on your iPad.
- Then, click on the **Create Link** button and send the link to your contacts to participate in the FaceTime call.
- After they've joined the group video call, simply swipe up and return to the home screen.
- Follow up by opening any of SharePlay compatible apps. For instance, Apple Music.
- Go ahead and select the content to use for SharePlay. Play a song, and you'll see the SharePlay button working in the FaceTime pop-up menu.

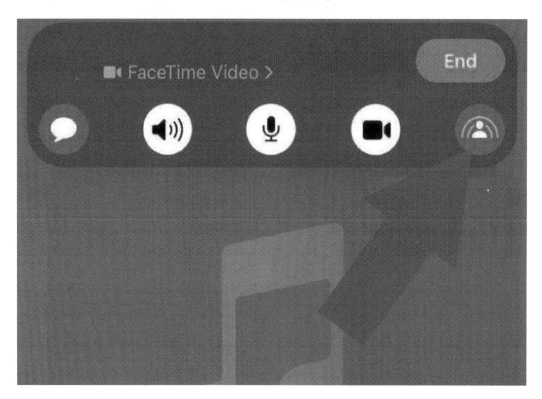

- The music will start playing simultaneously on the Apple devices of every participant. However, participants must be subscribed to the streaming service. Otherwise, they'll be prompted to navigate through the signup or free trial process.
- However, the playback controls will appear on the screen of every participant who has access. They can go ahead and tap Fast Forward, Play, Pause, or Rewind.

## Using SharePlay to Watch Videos Together

The illustration above shows how you'd use SharePlay to listen to audio alongside other participants in a FaceTime call. Now, you'll see how to do the same for videos.

- Start by opening FaceTime and creating a group call.

- The moment every participant has joined, launch any supported apps. For instance, the Apple TV app.
- Follow up by selecting the video that you wish to watch and clicking on the Play button.

## Turn On Portrait Mode for FaceTime

- Launch FaceTime.
- Access the Control Center on your iPad by swiping down from the upper-right corner of the display.
- Then, click on the **Video Effects** tile.

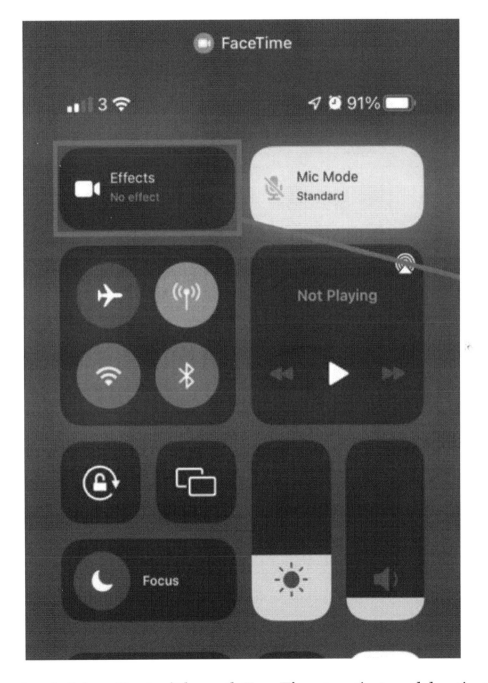

- Go ahead and click on **Portrait** beneath **FaceTime** to activate and deactivate the effect.

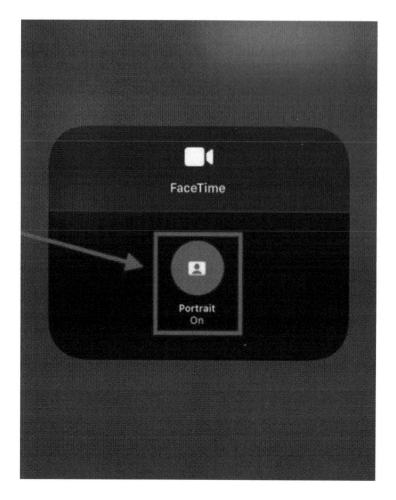

- This will ensure that your background is blurred before the call commences.

## Turn On Voice Isolation/Wide Spectrum for FaceTime Call

- Access the Control Center on your iPad while on a FaceTime video call.
- Next, click the **Mic Mode** button that appears on the right side of the display.
- Then, click on the **Voice Isolation** button (a checkmark will come up after selection) underneath the FaceTime column to reduce or restrict ambient noise.
- Then, click on anywhere on display to go back to the FaceTime window.
- Alternately, you can click and select the **Wide Spectrum** button to allow or improve the background noises.

## Turn On or Off Center Stage on iPad

Center Stage uses artificial intelligence to balance the front-facing ultra-wide camera when you use supported video apps such as FaceTime. Once you start moving around, Center Stage will help to keep you and everyone else within the frame.

- Launch the FaceTime app.
- Once the app launches, go ahead and swipe down from the upper right side to access the Control Center.
- Next, click on the **Effects** settings tile.

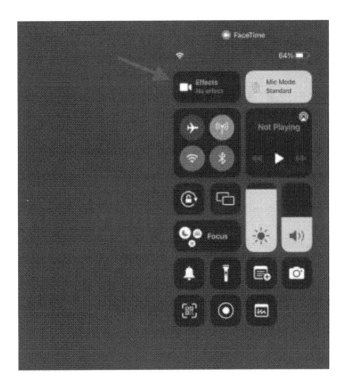

- Then, click on **Centre Stage** to turn it on.

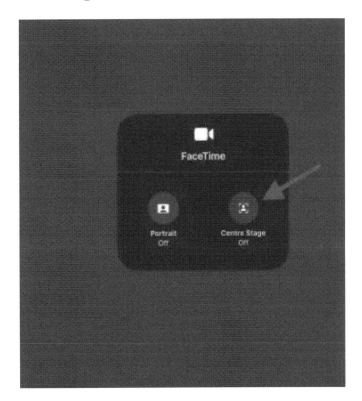

- That's it. The feature will start working while chatting with people via FaceTime. Go ahead and position your iPad anywhere and attempt to move back and forth in front of it. Your iPad's camera will then follow you from place to place.

# Imessage

How do I use iMessage?

Having an Apple ID on both your iPhone and any other iOS device, or have any Mac computer with OS X El Capitan or newer, then you can use iMessage for free. People using incompatible devices (e.g. Android) can also download the app from the App Store if they wish to send messages from their phones to others who are using Apple products.

What devices can I use iMessage with?

iMessage is only available to users who are logged into their Apple ID on their iPhone or iPad. However, you can use iMessage with a friend's device as long as they're also signed in to their Apple ID. If you're not sure whether your friend is using Apple products, then ask them—you don't need to be tech-savvy to differentiate between the logos!

If you've got an iPhone and OS X El Capitan or newer, you can use iMessage for free. Download the app from the App Store to start using iMessage.

How do I use iMessage?

Launch the Messages app and then follow these simple steps:

Tap the "i" icon in the top right corner of your screen. Tap Send Message to start a new conversation or Continue to continue an existing conversation. Select whom you would like to message. Type your message and send it!

Remember, if you have already got an active conversation, you can simply tap on it to pick up where you left off or tap Reply if you have a back-and-forth with someone. You can also make changes directly in your messages, such as moving them around by dragging them to a different spot in the thread.

How do I create an iMessage conversation?

If you are currently using a conversation, you can start a new one by tapping the "i" icon at the top right of your screen and selecting New Message. Next, type in whom you would like to message and then begin typing your message. You can always edit or change your message before sending it. You can also add people or groups to the same conversation until all of you are chatting away.

How do I use the predictive text feature for iMessage?

If you're having trouble typing on your iOS device due to its keyboard being too small, any mistakes that you make while typing will be corrected automatically by iOS' predictive text feature. If you're still having trouble typing on your device, you can adjust the text options to a larger default size by going to Settings > Display & Brightness > Text Size.

How do I stop someone from seeing my iMessage previews?

If you don't want someone to be able to see your incoming messages before you have read them, simply go to Settings > Messages. You can then turn off the Ask Before Deleting option so that messages will be permanently deleted from your device after being opened.

How do I turn on reading receipts for my iMessage conversations?

With this feature, you can see when someone has read your message. Read receipts are turned on by default, but you can disable them for future messages by going to Settings > Messages and then turning off the Send Read Receipts option.

## Send or Receive Money

Sending money with Apple Pay is simple. Simply click the FaceTime camera button in the lower left to get started (under "Send $XX.00 to <recipient's name>") or the Camera Roll icon in the lower right (under "Send $XX.00 to <recipient's name>"), take a photo of your credit card, enter your four-digit passcode on your device, touch Done, and then send money! This process takes just seconds—much faster than entering all of your credit card information manually!

Apple Pay makes it simple to receive money. Simply add funds to your Apple Pay account with a credit card or debit card, scan the QR code on the message, or enter your passcode, and then choose "Receive money" from the list of available options in Wallet.

If you need help adding funds to your Apple Pay account or integrating Apple Pay with iTunes, please visit: https://support.apple.com/en-us/HT204210.

If you need help with more advanced shopping, like setting up Apple Pay on a Mac and creating shopping lists, managing gift cards and loyalty accounts, adding custom payment information on receipts, and much more besides: https://support.apple.com/en-us/HT204200.

If you are not sure how to activate Apple Pay, simply enter your passcode and follow the on-screen prompts: https://support.apple.com/en-us/HT205463.

## Whatsapp

1) Downloading Whatsapp on iPad: If you haven't already done so, then find the Whatsapp app by searching for it from your Apple store or from the iTunes App Store and download it onto your tablet (whichever way you prefer). Then, follow the on-screen instructions to install Whatsapp.

2) Activating your Account: You'll need to activate your account before you are able to use it. This is accomplished by using Google Camera to scan the QR Code (the QR code will show a text that relates to your phone number). Once you have scanned the QR Code and your account has been activated, you'll now be able to start using Whatsapp.

3) Making Basic Calls with Whatsapp: This is pretty simple! Simply enter the phone number and press the "call" button. Your calls will be free via Wi-Fi and data charges are very low if you choose not to use the data option for making calls.

4) Making International Calls: If you want to make calls to any other country, then you will need to add a phone number from that country. On your iPad, you can accomplish this by going to the Settings tab in WhatsApp.

5) Using WhatsApp Web on your iPad: There are some very handy features that are only available through using the Whatsapp website – such as sending money (via PayPal), tipping people, saving photos and videos and sending images. The simple way to access these features is by going onto the web version of Whatsapp on your iPad, logging in with your Apple ID and password and then clicking through to the relevant feature (such as tapping on a photo or video and then hitting 'share').

## Creating Accounts in Social Media Networks

1)Open the Settings application, then select Social Media.

2) Select the account you want to set up, like Facebook or Twitter, and follow the prompts.

3) Enter your name and email for your new account, then type in a password for it.

4) Tap Trust this Computer and enter your lock screen passcode if you have one enabled on your IPAD device.

5) Choose if you want to allow push notifications from this account with Push Notification Apps.

6) Sign in with Facebook, Twitter or Google Account to sync contacts from that network with the contact information stored on iPad.

7) Press Next when you are finished, and a link to your new social media account will appear under Accounts.

8) Tap on the new link and enter your login credentials for this social media account.

9) Press Done, and then tap on your name to update it using the options provided.

10) Select one of the suggested contacts to add as your profile picture by tapping Choose Photo, or tap Choose Contact to choose an existing photo from an address book contact.

11) You can now move your cursor over the new icon for this social network account located on the left side of your screen, and then tap Edit Settings in order to provide more information about yourself.

12) After signing into your new account, you will need to download your contact list for it. Tap Download Contact File, and then tap Save.

13) Enter a name for the file and tap Save to create a backup copy of your new social network contact list. Tap Done in order to save the changes that you have made.

14) To return to the previous screen, tap back until you reach your profile tab, and then tap Edit Profile.

15) Once back on this screen, tap Done in order to update any changes made to your information.

16) You can also edit certain elements of your profile here, like the text under About Me or remove information from it by tapping Edit Info.

17) Tap Done and then tap your name to update it once again.

# Chapter 5:

# Maps, notes and utilities

## MAPS

You can see your location in the Maps application and zoom in to view the info you want.

### Allow the maps app to make use of Location services

To search for your location & get precise directions, your tablet has to be connected to the internet & Location Services need to be Enabled

If Maps shows a notification that Location Services has not been activated, touch the message, touch Turn On in the Settings application and then activate Location Services.

### See your current location
- Touch the Track icon⌖.

Your location is marked in the center of the map. The top of the map is north. To display where you are heading at the top instead of north, just touch the track icon ◀. To display north as the top again, touch⌖ or ⊙

### Choose the appropriate map

The buttons at the upper right side of a map would signify whether the map you are in is for driving🚗, viewing from a satellite🌐, exploring🏬, or riding transit🚊. To pick another map, adhere to the directions below:
- Touch the button at the upper right part of your display.
- Pick another type of map, and then touch the Close icon✕

### View a map in 3D

Drag 2 fingers up on a 2D map. (When using the satellite map, touch the 3D icon close to the upper right part of the map.)

On a 3D map, you can do any of the below:
- Change the angle by dragging two fingers down or up.
- Zoom in to see buildings & some other features in 3D
- Go back to the 2D map: touch 2D close to the top right part of the map.

# Notes

You can use the Notes application to quickly write down thoughts or arrange detailed info with checklists, pictures, Internet links, scanned docs, handwritten notes, & sketches.

## Create & Format a note

- Click on the new Note icon ⊡, then type what you want.
- Click on the format icon Aa to make changes to the format.
- Click on the **Done** button to save your note.

Touch the Checklist icon ☷ to add a checklist to your note. When you are done with an item, touch return to enter the next item.

Touch the Table icon ⊞ to add a table to your note,

## Draw or write in a note

- Use your Apple Pencil to draw and write your note. You can also tap on the Manuscript icon Ⓐ to write or draw with your finger
- Change materials or colours: Use the Mark-up tools.
- To change your handwritten text to typed text as you write with your Pencil: Touch the handwriting tool (on the left side of the pen) and then you can start writing.

**Add videos or photos**
- In your note, touch the Camera icon.
- Take a picture or a video, or pick one out of your photo collection.
- To edit attachments' preview size, long-press the attachment, and then click on Large images or Small images.

To draw on a picture, touch the picture, and then touch the Markup icon.

To save videos & pictures taken in the Notes application to the Photos application, open the Settings application, touch Notes, and enable **Save to Photos**.

**How to scan a document**
- In your note, touch the Camera icon, and then scan the Docs.
- Arrange your device in a way that the document appears clearly on the screen; your device would automatically fetch the page.

To snap the page by hand, touch the "Take Picture" icon.

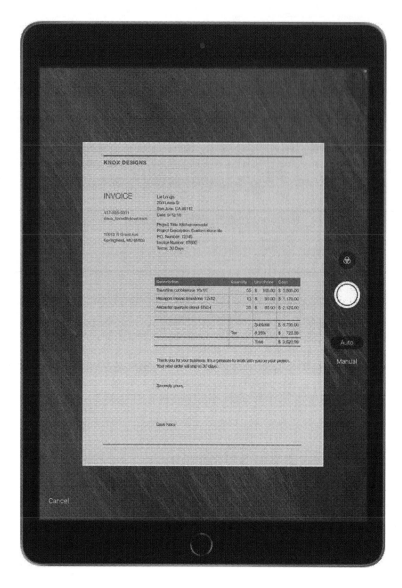

- Scan more pages, and then touch the **Save** button when you are done.

## Create quick notes anywhere on your device

You can utilize the **Quick Notes** feature to write down information over any application or screen on your tablet. You can access your Quick Notes in the Notes app.

### Creating a brief note
1. Do one of the below:
➢ Open the Controls Centre, and touch the Quick note icon.

(If you can't find the Quick Note icon in the Controls Centre, you can add it in the Setting application> Control Centre, then select Quick Note).

➢ Swipe up with your Apple Pencil or your finger from the bottom-right corner of your screen. Do any of the below:
- Begin a new note: click on the compose icon
- Enter text: Write or type using the Apple Pencil.
- Add link: click on Add Link.
- Swipe right or left to access your other Quick Notes.

### To view & organize your Quick notes
You will find your quick notes in the Quick Notes folder in the Notes application.

### Lock notes on your device

You can protect sensitive information by password-protecting your notes. The Notes application utilizes one passcode for all locked notes. You can also utilize Touch ID as a means to access your locked notes, but don't rely on Touch ID as the only means to access your notes.

### Setup your notes password
- Head over to the Settings application, touch Notes, then tap on Password.
- If you have many accounts, select the account you wish to set the password for.
- Enter your password as well as a hint to aid in remembering the password.

You can also enable Touch ID.

### Lock a note
You are only allowed to lock notes on iCloud & on your device. You cannot lock a note that has Pages, PDFs, Keynote, video, audio, or Number documents in it.
- Click the More Options icon after opening the note (in the upper right of your display).
- Touch Lock

Click the More options icon in the top right corner of your screen, then tap the Remove button to unlock a message.

# Create Reminders

An important part of your day-to-day life is your ability to stay on top of all the things you need and want to do. This can be a lot more difficult than it sounds when you don't really know how much time you have before an appointment, what tasks must be completed before leaving work, or which bills need to go in the mail. Despite this, it's easy to wrack up quite a few reminders on your iOS and Android devices.

Reminders for iOS and Android devices share a lot in common but also have their own unique features to make it easier for you to get things done. This article will go over some of the different types of reminders that are available, who can create them, how they function, and how they can be used in your daily life.

# Select & Translate Live Text

International readers may find the iPad user guide on Apple's website helpful. The following is a brief look at just one of the many features offered: Select & Translate Live Text.

If you use your iPad in another language, this guide will help you get the most out of it. In addition to basic setup information, you'll also find tips for editing and revising documents, time and date formats, and international temps and measurements, among others.

For the Spanish reader, for instance, you'll find a translation for "offline" as sin conexión and tips on navigating through folders and documents.

# Chapter 6:

# Camera and photos

## *Using the Camera app*

Take photos and videos on the iPad using the camera app. It is designed to work closely with a Photo app that allows you to view, organize and edit your pictures and videos.

To use the camera from the lock screen, follow these steps:

You can quickly access the camera from the lock screen. Simply swipe left to open the app immediately. You do not need to open the device first.

## *Zoom in and out*

When taking a picture, pinch to zoom in and out. Keep in mind, however, that the iPad employs digital zoom rather than optical zoom (as seen on larger cameras). Digital zoom tries to mimic optical zoom. As a result, zooming leads to poor image quality.

## *Camera direction*

Rotate the iPad to change camera orientation from landscape to portrait.

## *Turn the camera over*

Tap the Flip Camera button in the upper right corner to switch between the front and rear cameras. You can photograph yourself using the front of the camera. Also, use the front camera for video conferencing applications such as FaceTime.

## PHOTO

Please take a photo or video on your iPad, and you get it directly in the Photos app. You can use photographs to make slideshows, share media in a variety of ways, and organize your photos and videos in albums.

## *Image processing*

The Photos app also allows you to edit your photos with simple options like rotating, enlarging, removing red-eye and cropping. This is an easy way to fix small image problems without the use of a sophisticated image printer.

To access the options, tap the Edit button while viewing the image on full screen.

Editing options appear next to the screen. When you're done editing the image, click Done.

## Share Photos

There are many ways to share pictures from your iPad. For example, you can send a photo as an email attachment, post it on Facebook or Instagram, or play it in a slideshow on your iPad. Simply tap the Share button in the upper right corner, select the photos you want to share and then select the option you want.

You can transfer photos from your iPad to your computer by connecting your iPad to your computer using the included USB cable.

**Method 1: Import to Mac**
- You're going to need a USB cable. First, connect your iPad and Mac computer using the USB Cable.
- Open the Photos app on your computer.
- The Photos application will display an import screen containing all the images and videos on the connected device. Click the device name in the image bar if the import screen does not appear automatically.
- When prompted, use your password to unlock your iOS or iPad OS device. If your iOS or iPadOS device asks you to trust this computer, tap Trust to continue.
- Select a location to import your photos. Next to Import, you can select an existing album or create a new one.
- Choose the pictures you wish to import and tap **Import Selected**, or click **Import All New Photos**. Before removing the device from your Mac, let the operation finish

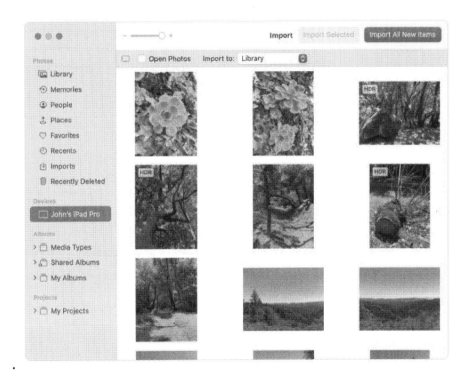

**Method 2: How to download photos from iPad to Mac using iPhoto**
You can also use iPhoto to download photos from iPad to Mac. Follow the steps below.

**Step 1:** Connect the USB cable to connect your iPad to your Mac.

**Step 2:** Open the iPhoto application on your Mac. iPhoto shows the photos saved on your iPad.

**Step 3:** Choose which images to import. Then, choose Import Options.

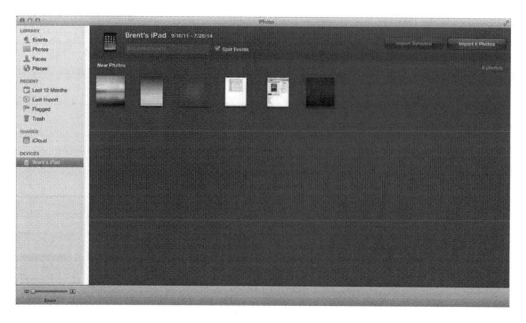

**Step 4**: When done, you will be asked if you want to delete or keep the images after importing.

**Method 3: How to copy iPad photos to Mac with Image capture**
The steps below will demonstrate how to transfer iPad photos to Mac using photo capture.

Step 1: Connect your iPad and Mac computer using the USB Cable.

Step 2: Open the Image Capture app on your Mac.

Step 3: Choose the photos you wish to import to your Mac.

Step 4: Select a location to save your Mac photos. Then click Import or Import All.

Step 5: When you are done, you will see that the imported images have a green check mark.

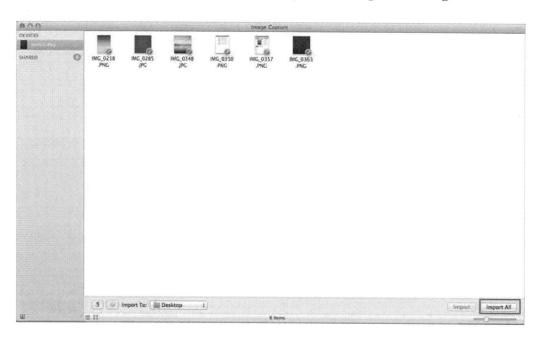

# Sir and Apple Music

The most crucial fact about Apple Music is that nothing actually streams via it. Instead, it downloads all of its content on your device before playing back using High-Efficiency AAC streaming at 256kbps (your mileage may vary). This is important for your storage space and for your data usage on LTE networks. If you're on a high-speed WiFi network, you will have access to music without downloading anything from Apple's servers.

For users with large collections of their own music that they want to integrate into their listening experiences, Apple Music has created a folder called "iTunes in the Cloud," which appears in the Music app and in iTunes on the desktop. This can be used as a new iTunes library for organizing and accessing all of your local music alongside the Apple Music library. Yes, this includes any DRM-free AAC files you purchased from iTunes or ripped from CDs.

The biggest feature of Apple Music is Siri integration. To play songs from iTunes Radio, use Siri, Apple Music or Beats 1, and she will do her best to find them for you. You can also cast music from the cloud to an AirPlay speaker.

Once you begin using Apple Music, you may access your music library in a variety of various ways: Control Center > Music. This is also where you can seek out new content from the global iTunes library, which has been increased over time to include all of your iTunes purchases and rentals as well as music that's downloaded from the iCloud Music Library.

# How to listen to Music

iPad has a powerful music library. You can listen to iTunes music, podcasts, videos and other apps' sounds.

This is an introduction to how to use the iPad such that you can listen to music.

Click on the "Music" app for this device, then click on "Playlists". Do the same for "My Music".

Click on "Library" and then "Playlists."

From the items in the playlists you have not added, select the playlist that you want to listen to. After this, tap "Play" to start listening.

The playback of songs and videos keeps going as long as you keep your iPad connected. You can listen using any app like iTunes, internet radio or podcast player by selecting the playlist's name from there. Note that if an audio file is too large, it can take time before playback starts after selecting it in the app library.

If you want to listen to music as background sound, you can use the iPad in multitasking mode by selecting the "play" icon on the home screen.

The iPad's Home button has functions that are different depending on what app is running. The following shows how to use the button when listening to music or watching a video. You can pause or resume playback or fast forward and reverse by tapping once. You can return to the app menu or go back to Home Screen by double-tapping

To fast forward, double-tap and hold down while moving two fingers in the direction you want to fast-forward (to move more quickly). To reverse, double-tap and hold down while moving two fingers in the direction you want to reverse (to move more quickly). You can also move with your fingers on the screen without tapping.

To play music or videos in album format, go to "Album List" by tapping "Library".

To play music or videos in a list, go to "Artists", "Albums", or "Songs".

Tapping the top bar allows you to move the current song's position back to the beginning of another song. *You can also tap and drag songs to rearrange them on your playlist.

Click on an album cover for that tune's information page.

## Apple Music Settings

Users can access any song in the iTunes catalogue through Apple Music, the company's music streaming service. The service comes with a free trial period, which allows users to test the service and find out if it is a good choice for them.

Starting Apple Music on an iPhone

To get started with Apple Music on an iPhone, you need to open up the Settings menu. This can be done by tapping on the settings icon on your home screen or by swiping down from the top of any screen on your iPhone. Once you are in Settings, tap on iTunes & App Store.

In the iTunes & App Store menu, tap on Apple Music. This will open up a list of options that let you set your preferences for Apple Music. The first option is to sign up for the free three-month trial. After that, you can pay $9.99 per month to continue enjoying the service. If you do want to continue using Apple Music after your free trial expires, you have to tap on the Become a Member button and then tap on Subscribe for $9.99 per Month.

Once you have signed into Apple Music, a new screen will appear with all of the different categories of music that are available through the service that you can search through and listen to at your leisure. You can find and play your favourite music, albums, new releases, playlists and artists on the Apple Music app.

Starting the Apple Music app

When you tap on the Apple Music tab in the settings menu, you will be brought to a page with a few options at the bottom of your screen. Most of these options are tutorials that show how to use different features in the app, such as adding and removing songs to playlists. However, there is also a search bar where you can enter specific artists or genres that you want to listen to. Once you do find something that you want to listen to, tap on it and it will begin playing right away.

By offering users the option to listen to their favourite music in various ways, Apple Music is a free way for app users to enjoy all of their music. Learning how to set up and use the Apple Music app will give you a quick and easy way to listen to songs that you have downloaded onto your iPhone.

## Watch TV and Movies

Best apps for watching TV and movies on your iPad.

- YouTube app: This is a great app for watching any type of video without buying anything. However, with ads popping up every 20 seconds, it can get annoying if you're not careful with what videos you view. You can pay $9.99 for the ad-free version.

- CinemaNow: This is an app that allows you to rent movies online and watch them on your iPad. It has a lot of different movies and TV shows, but I've found some good ones to watch at the least are "Dinner For Schmucks", "Minority Report" and the "Harry Potter movies". You can purchase the first three movies in this list if you like, but it's not necessary. All of these apps are only $2.99 a month.

- Vudu: This is another great movie rental app that's extremely similar to CinemaNow. Their movies are a bit cheaper in price, but you'll have to purchase them at $7-$10.

- PopcornFlix: This app is a great choice for anyone who likes to watch classic movies. It does not have any new movies or TV shows, but it does show oldies such as "Frankenstein", "Dracula", (the original) "Planet of the Apes" or "The Wolfman" which is pretty cheap at $2.99 a month.

- Netflix: This app is free and it's also where I watch my favourite TV shows like "The Walking Dead", "American Horror Story", "Castle" and more! You can pay $8.99 a month for the Netflix + HD plan, which will let you watch videos in 'HD'.

- EPIX: I have not heard of this app before, but it's pretty much just another movie rental app. You can purchase movies here if you'd like as well.

- Showtime Anytime: This app is only $11.99 a month and it's great for watching TV shows, especially "Dexter", "Californication", or any other Showtime original show. It also gives you access to movie trailers, full episodes and more!

- HBO GO: The great thing about this app is that you already have it if you have an HBO subscription from cable TV. Just buy the HBO Horizon plan for $14.99 a month, which includes HD access.

## Podcast on your iPad

1. Create a playlist in iTunes on your computer first before importing the podcast.

2. Open up the Music app on your iPad and find the "Podcasts" section of your library. Then tap "Subscribe to Podcasts" (in blue) and press enter twice. From there, you'll be able to scroll through the list of available podcasts and choose this one!

3. On your iPad:

a. Tap the Podcasts icon in the Library.

b. Tap "Subscribe" to begin listening.

c. Tap Next in the upper right-hand corner of your iPad. You'll be asked to enter your Apple ID or create a new one if you haven't already done so!

d. Enter your password and tap Next again when requested to do so. Note that you won't be able to log into this app until you've entered this information correctly!

e. Once you've created an account, tap OK and then install the podcast via "Playlist" or "Podcasts.

f. Tap on the podcast to open it.

g. Tap the play button (third icon from the right) and then press "Done" when asked if you're sure! If you're having trouble finding the podcast, tap "Search" in the upper right-hand corner and enter your search term in there (e.g. "IPADUSERGUIDE") - but note that this could take some time to work!

h. Tap "Play" once the episode has finished downloading and it will begin playing automatically on your iPad! It's that easy!

4. On your iPad:

a. Go back to your iPad's Podcasts app and tap "Done" to exit out of this app after playback.

b. Tap the Podcasts icon in the Library and tap "Playlists" (fourth icon from the right) and then tap "Add Playlist".

5. Here, you should be able to drag & drop the podcast into your new playlist!

6. Once you've got this set up, you'll want to create some smart playlists/playlists with smart rules that can automatically keep tabs on what's going on with your podcasts as they are released.

# Chapter 7:

# Ipad accessories

## Airpods and Earpods

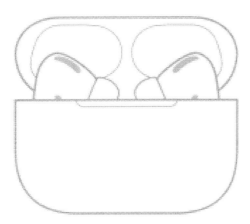

How to pair Airpods Headphones and mobile device:

1) On your mobile device, open the Settings menu and tap "Bluetooth Devices." Press the Airpod button on one of the Airpods. (If it doesn't light up or blinks once, check to see if your battery is charged fully.) 2) When pairing is complete, you will see a message saying "Pairing Device" in the top bar of the Bluetooth settings. 3) On the left side of the Bluetooth settings, you will see your Airpod on this list.

2. Use your Airpods to make a phone call:

4) On your mobile device, go to your call screen (phone or FaceTime screen). Press the small circular button on one of your Airpods, and wait for the tone. 5) When you hear the tone, speak into it to take the call.

3. Use your Airpods with Siri:

6) You can also talk directly into one of your Airpod earpieces to use Siri or play music using voice commands. Simply press and hold the circular button on one of the Airpods. 7) To end a call, either tap the same button you used to start the call or double-tap one of your Airpods in your ears.

3. Use your wireless Airpods with other devices like iPhones, iPad and Mac:

8) You can also use your wireless Airpods with other Apple devices like an iPhone, iPad and Mac. Simply open Settings on any device that you want to pair Airpods with, then click Bluetooth. 9) Click

the "i" icon next to the name of your products and it will open a new screen with details on them (make sure they are in range).

3. Set up your Airpods on your Mac:

10) Open System Preferences and select "Bluetooth."To add your Airpods, click the plus sign in the top right corner.

11) Select one of your Airpods and a new screen will appear. Select "Connect" underneath the device name. 12) You will see a message that says devices have been paired successfully when they are connected or turn off Bluetooth to disconnect devices if you want to stop using them at any time.

13) Now that you know how to pair and connect Airpods, make sure you have enough battery life by clicking the circular button on one of them.

4. How to charge Airpods?

14) After you have made the appropriate connections on your mobile device, you can now charge your Airpods. Simply take off the case that comes with Airpods and put it in the case that comes with them (the case is meant to hold a single earbud either in your ear or on a cable).

15) Once you connect the case to a power source, you will see the light on one of your earbuds turn on, indicating them charging and a battery indicator will appear in the Settings menu next to "AirPods." This can take up to two hours, depending on how much battery life they have left.

5. How to customize your Airpods wireless earbuds

16) Change the size of your wireless earbuds: You can change how long and how large you want your Airpods to be by moving them up or down in the case that comes with them. 17) Switch between the different sizes of headphones: To switch from one size to another, simply push on the left Airpod's side so that it clicks into place and then reposition it. To remove one, click on it and pull out.

18) Adjust the fit: You can also adjust the fit of your Airpods in order for them to stay in place, simply pull up at the part of your earbud closest to your face. If you are having trouble removing or putting in one of your earbuds, you might need to adjust how far they go into your ears.

6. How to take care of your Airpods wireless earbuds

20) Cleaning your Airpods with a wipe: You can clean both microfiber and rubber each with a soft, damp cloth. If you want even more cleaning power, use a non-abrasive soap—especially if you have been sweating while wearing them.

21) Know that you can't use the Airpods with other headphones: The Airpods are not compatible with any other headphones. If you want to use headphones, you will have to take off your Airpods.

7. How to pair and connect your wireless Airpods earphones with multiple devices:

22) Pairing and connecting your Apple devices with multiple pairs of Airpods: If you want to connect multiple pairs of Airpods in case you lose one or if you want to share them with someone else, follow these steps (make sure the other person has his or her own mobile device already set up). 23) Make sure you have the latest version of iOS on your mobile device. If you don't, you will need to update it in order to connect your Airpods to multiple devices.

24) Go to Settings and tap Bluetooth. Once the Bluetooth menu appears on your Airpods, tap the name of the Airpods in that list. 25) You will see a screen that says "Enter Passcode" with a six-digit code next to it. 26) On the mobile device you want to connect, hold down one of your earbuds until it chirps and then enter that passcode into the Enter Passcode screen by tapping its numbers on the mobile device's display.

8 . How to use your AirPods as a headset

27) After you enter the code, wait for it to say that they are connected. Your Apple devices will be able to detect which are the left and right earbuds so that you can use them with voice commands or tap on one of them for Siri, for example.

9 . How to use Siri with your Airpods?

28) Make sure you have all your settings set up on your mobile device so you can use Siri with your Airpods. Here's how:

29) Open Settings and select the "Siri" tab under "My iPhone. " 30) Tap "Allow" to turn on Siri. 31) In the Settings menu, select "ALL" and then tap "Allow" to turn on Siri. Don't worry—you can turn her off at any time if you want.

10 . How to use AirPods with iMessage

32) Press and hold the circular button on one of your earbuds until it begins making a sound and then release it when you see the blue light begin flashing. 33) Scan your mobile device's number by connecting your Airpods to it with Bluetooth (see above for details). 34) The earbud will vibrate for about a second when connected so that you know you have successfully paired.

Certain models of iPads come equipped with an in-built speaker system, but for those without one, earpods are a must! They're also a great way to privately listen to music or audiobooks on your device when others want some peace and quiet as well.

You can use earpods to make/answer phone calls too.

Locate your earpods and make sure they're turned off. Find your earpods and make sure they are disconnected. Hold down the centre control button on each earphone to accomplish this - the one that looks like a play/pause symbol - until the LED light on both earpieces changes from green to red. From now on, when you want to turn them off, just long-press this same button again until the LED lights go out.

To make or receive a phone call from your iPad, you'll need to connect the earpods to your iPad by sliding them into the top holes on either side of the device. If you have an iPhone 6+, then there's no need for this step. Simply make/receive calls using the in-built speaker system on your device.

To quickly return to your media playlists, albums or audiobooks, just tap and hold a control button on either earpiece - the one that looks like a playback/stop symbol - until all four LEDs light up, indicating they're active.

To pause or rewind, just tap the play/pause button on either earpiece and hold down until it turns green, indicating it's active.

To increase or decrease the volume of your media, just tap the control button that looks like a small "i" on each earpiece and holds it down until it changes colour - from green to blue to red - indicating that volume has been increased or decreased.

To clear your earpods of any audio files you've played, just double-tap either side of the right/left control buttons (the one that looks like a play symbol) until they're both blue. This cancels all audio playback.

You can also use your earpods as an audio output device for photos, podcasts or audiobooks. To do this, simply plug them into your dock connector and place them in the headphone jack on the back of your iPad. The front two holes will provide a standard 3.5mm headphone jack so you can use all of your existing devices with it as well, including any other headphones you've got lying around.

Advanced Features:

Turn on/off the iPad's built-in microphone using the volume buttons on either side of the earpods - these control the LED lights on both sides of them to indicate when they're active or passive.

# Apple Pencil

The Apple Pencil is a stylus that works with the iPad Pro, which allows you to draw, take notes, mark up documents and much more. The Apple Pencil is a powerful tool that can help turn your tablet into a high-quality drawing pad. The pencil communicates directly with the processor of the iPad via Bluetooth 4.0, so you do not need to charge it separately during use.

People use the iPad for many different things like reading emails or playing games. While that may be all you want to do, the iPad Pro is much more capable and can perform much more than just what is on the surface. The Apple Pencil works with a variety of apps on your iPad Pro including

Apple Notes, Pages, Numbers, Keynote and other apps in iWork as well as third-party options from the App Store.

Connecting the Pencil to the iPad Pro

You need to connect the Apple Pencil to the iPad Pro for it to work with your tablet. This is done by removing the cap and holding the Pencil while touching the charging port on your iPad Pro. The LED light on the Pencil will blink blue, indicating it is charging, and then when red, it will begin to charge.

After connecting to your iPad Pro, you can begin to use your Apple Pencil. In this section, we are going to show you how to connect it with a variety of apps.

The first step is to open Pages and erase any previous notes, which can be done using the "New" button on the bottom left corner of your page.

Next, select the Apple Pencil from the list of recently used tools, and then click either the "Insert" or "Draw" tool. You will see a screen appear where you can set your options for drawing.

When setting options for an app, you have to make sure you are using the correct settings for that app. For example, if you are using Pages for your drawing layout, then in this example, we will use "Baseline Grid". The "Line Colour" option is not specified because it uses the colour of your finger when drawn on a page, while applications like Numbers use the colour of ink on paper.

If you are using the "Draw" tool, then choosing "Colour" will allow you to choose from colours on the colour wheel. However, if you are using the "Insert" tool, then you can choose a new object in the App store and then insert it into your document.

When drawing or typing on your document, there is a marker that appears when holding your Apple Pencil over text. This marker will allow you to select, cut, copy and paste as well as move objects around in a document. When moving objects, just hold down your finger and slide it wherever you want. If you are copying or cutting text, it will appear in the separate menu bar at the top of an app window.

When selecting the text, you will notice there is a small popup that appears when holding your Apple Pencil over the text. This will indicate the size of selected words and phrases. The colour of the popup will indicate what the selected word or phrase is. For example, blue means it's not a proper noun or name. Red means it's a proper noun and green means it's a name.

Not only can you do all this with your Apple Pencil and iPad Pro, but you can also zoom in up to 2x Zoom by double tapping with two fingers, as well as accessing other features via a 3D touch from the home screen icon or most used applications page inside Settings app. You can access these settings by holding the Apple Pencil on the Settings app and then selecting "More" and then "General".

While drawing with your Apple Pencil, you are using what's called a "Tip Pressure Sensing module", which allows for 256 levels of pressure. This is how you can create more fine detailed drawings and get a much more natural feeling when using your fingers to draw on the screen.

The Apple Pencil also has an "eraser" built in that allows you to choose what type of eraser you want, for example, if you want it to be sharp or soft. You can change this from the drawing options menu in each app.

If you want to delete a line or object, then you can use the erase function in your app. For example, in Pages, you can press "Delete" once and it will clear away the selected area. If you hold down your finger on the eraser for a few seconds, it will become very sharp, then remove that same area completely from your document.

If you are working on a tablet like an iPad Pro where there is no hard keyboard and using only one hand, then the Apple Pencil is a great aid to help draw things and write. Such as if you are writing a shopping list or planning your family vacation. But it also has its advantages when working on a Mac as well. If you're using an app like Keynote and want to make a quick change, then you can use the Apple Pencil to quickly write over the text and make a new change.

## How to use Scribble

You must download Scribble as soon as possible. Once it's installed on your IPAD, open it up and hit "sign-in" in order to sign into Scribble with an account. From here, select "create new" or "sign-in" if you have one already created. In the top-right corner of the interface, you can see the tabs which are available in Scribble. All of these tabs are used to help you create your art.

After you've finished signing in and have selected your artwork, select "Create" on any of the templates you have available. This will prompt you for your name, so select it and hit "continue". The next screen will ask for an image, so select an image from your device's camera or from Google Images (if you don't have an image already available, then enter a blank one). You'll also be asked to add tags before selecting "Continue".

After hitting "continue" on the previous screen, you'll be asked to select a colour palette. You can either tap "add a new one" or scroll through the provided options. When you've selected a colour, hit "add" and then "done" when you're finished picking your colours. Before parting with your creation, check for any mistakes by selecting "view image" or tapping on the "i" icon at the bottom of the screen. You can also select "share" if you're done adding to and editing your photo.

# Quick Note

Create Quick Notes anywhere on your tablet

The Quick Notes feature can help you to write information on any application or screen on your iPad. You can access all your Quick Notes in the Notes application.

Create a Quick Note

You can start a Quick Note in any of the ways below:

Using your finger or Apple Pencil, swipe up from the lower right corner of your screen.

Open the Controls Centre and click the Quick Notes icon.

(If you do not see the Quick Notes icon in the Controls Centre, you can add it in the Settings application> Controls Center, then select Quick Note).

Do any of the below:

Write with your Apple Pencil or onscreen keyboard.

Tap on Add Link to add a link.

Begin a new note: Touch the New Note icon.

Go to another Quick Notes: Swipe to the right or left.

View and organize your quick notes

Click the Quick Notes folder to see all your Quick Notes in the Notes application.

Note: You cannot lock a Quick note unless you move it to another folder.

# Chapter 8:

# Ipad security

## Two-Factor Authentication

When you enable two-factor authentication, Apple will text you a temporary six-digit code that you use as well as your Apple ID and password when you sign in from a new device. This will help significantly in preventing unauthorized access to your valuable data.

With two-factor authentication, you can only access your account on devices you trust, like your iPhone, iPad, or computer. Your six-digit verification code, which will be displayed automatically on your trusted devices, and your Apple password must both be entered if you have a new device, such as a computer, iPad, or iPhone, in order to access your account for the first time. By entering this code, you verify that you trust the new device.

For example, if you have an iPad and you're signing into your Apple account for the first time on your computer, you'll be required to enter your password and the verification code that's automatically displayed on your iPad.

Your password on its own is no longer enough to access your account. Two-factor authentication markedly improves the security of your Apple ID and all the personal information you store with Apple.

However, if you completely log out of your Apple account, erase the device and delete the data, or need to change your password for security reasons. When you sign in on the web, you have the option of trusting your browser, which means you won't be asked for a verification code the next time you sign in from that computer.

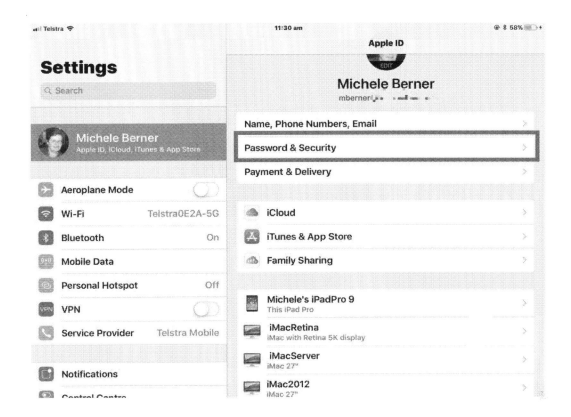

# Turn on two-factor authentication

Select **Password & Security**. Since I've already switched, it says On for Two-Factor Authentication. If your screen also shows On, you're good, and you can stop now.

Tap "**Turn On Two-Factor Authentication**." Then, tap on "**Use Two-Factor Authentication**" after reading through the information on how to use this feature. You have the option to use either a phone number which is linked to your account (this is automatically detected) or enter a new phone number that you wish to use instead.

Verification codes will be sent to the number you register with, and you will be given the option of receiving them via text message or automated phone call. Once you set up Two-Factor authentication, enter the password for your Apple ID, followed by a six-digit code. The two-factor authentication feature will be fully enabled upon successful input of both. For more detailed information, see Apple's support page, Two-factor authentication for Apple ID.

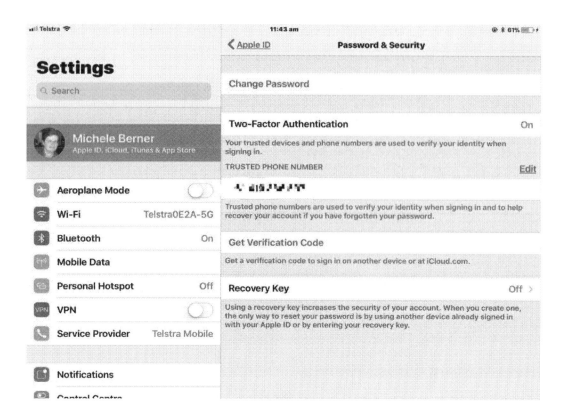

Here's an idea of what two-factor authentication looks like when I sign into iCloud.com on my Mac for the first time. I need to enter a verification code which I elected to be sent to my iPhone. Once the code is entered, I trust the device. Now I can login to iCloud.com from this Mac computer as it's been verified – it's been trusted.

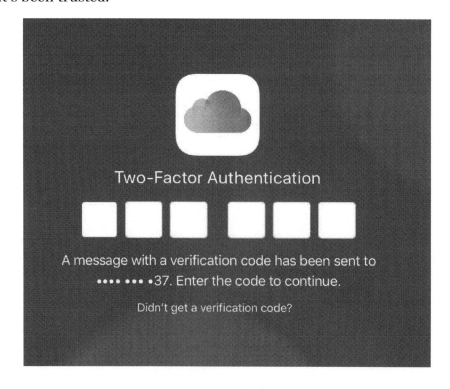

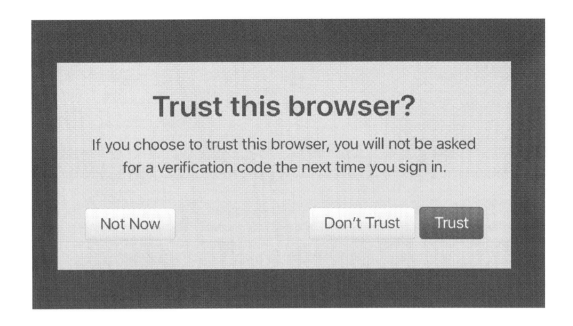

## Turn off two-factor authentication

On a computer, go to iCloud.com. Because two-factor authentication has been enabled, iCloud will ask you for a six-digit verification code before it allows you to sign in. So, approve your login by tapping on "Allow" on the request prompt to get the code. Then type it into the iCloud page in your computer's browser. Then click on "Trust" in the prompt to complete your login and finally bring you to your iCloud account's main page.

1. Inside your iCloud account page, click on **Settings.**

2. Click Manage - under the Apple ID at the top. You might have to verify yourself again.

3. You'll then see all of your Apple account settings. Click Edit in the security tab.

4. Two-factor uthentication will say On. Click "Turn off Two-Factor Authentication". In the pop-up box, confirm your choice.

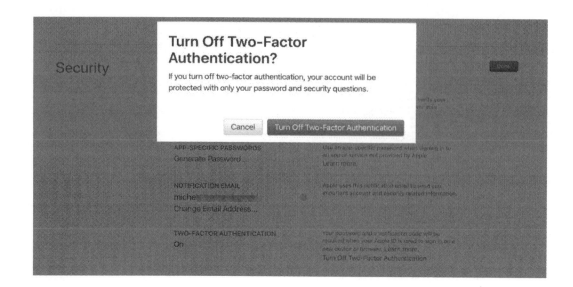

Security

**Turn Off Two-Factor Authentication?**

If you turn off two-factor authentication, your account will be protected with only your password and security questions.

Cancel     Turn Off Two-Factor Authentication

APP-SPECIFIC PASSWORDS
Generate Password...

Use an app-specific password when signing in to an app or service not provided by Apple.
Learn more.

NOTIFICATION EMAIL
miche...
Change Email Address...

Apple uses this notification email to send you important account and security related information.

TWO-FACTOR AUTHENTICATION
On

Your password and a verification code will be required when your Apple ID is used to sign in on a new device or browser. Learn more.
Turn Off Two-Factor Authentication

# Chapter 9:

# Setting and troubleshooting

## Secure your passwords with iCloud Keychain

## What is iCloud Keychain?

iCloud Keychain creates, stores and allows you to access all your passwords. It allows you to create more complex passwords, logins and not just use the same password for everything. It's built into all Mac and iOS devices, so it will if you have an iPhone, an iPad and a Mac computer. It is an easy method to create and manage all your passwords. You could also have a secondary password manager in case something goes wrong, like 1Password or Lastpass. You can also add your credit card details and personal information like your name, address and phone number and have them auto-filled whenever you need them.

## Enable iCloud Keychain

To enable it, go to **Settings**. Tap your **Apple ID banner**, then tap **iCloud.**

Scroll down and tap **Keychain,** and toggle it **On.** If prompted, enter your Apple ID password.

## How to create a randomly generated password using iCloud Keychain

You should never use the same password for more than one website login.

Open Safari and navigate to the website for which you want to create a login.

Select the password field in the form. If the website permits, Safari will select a strong password. Tap, **Use Strong Password** to use it, or tap, **Select my own password** to create your own. It might also offer a suggested password on the top of the keyboard.

If you don't see any suggested passwords, that site may have disabled iCloud Keychain for security reasons. You'll have to think up your own strong, unique password.

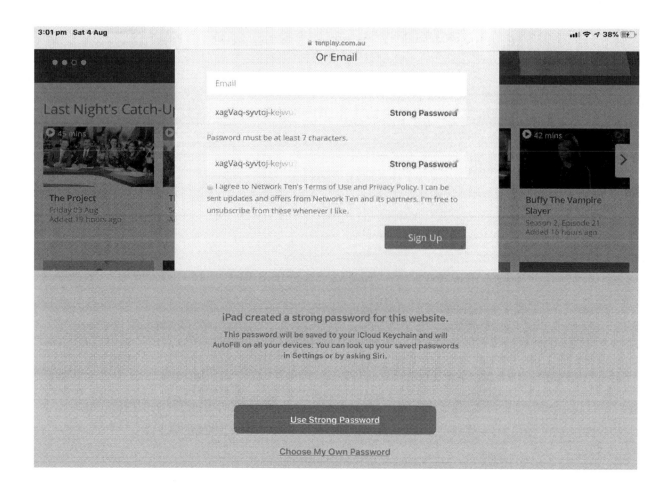

On some websites, when you tap in the password field, you'll see **Passwords** on the top of the keyboard. Tap that, and it will ask you to **Suggest New Password**, or you can tap **Other Passwords**, which will take you to your saved passwords in iCloud Keychain.

# How to access passwords using iCloud Keychain

The purpose of iCloud Keychain is to make it easier and safer to fill out passwords when you're using Safari. However, sometimes you may need a password and you're not trying to login into a website. If you're not able to use autofill to fill out a password, you can still access the feature. Your random-generated passwords are also stored here and can be accessed manually.

Go to **Settings>Passwords & Accounts**. Tap **Website and App Passwords**. Use Touch ID or your passcode if prompted to see your passwords.

Tap the login details for the website you want the password for.

Long-press on the password and tap **Copy.**

# How to delete passwords from iCloud Keychain

You may have stopped using an app or no longer want to visit a website. You can delete this information, so it is no longer stored in your Keychain.

Go to **Settings>Passwords & Accounts**. Tap **Website and App Passwords**. Use Touch ID or your passcode if prompted to see your passwords.

Tap the login details for the website you want the password for.

Tap **Edit.** Tap the website/password to select it, and then tap **Delete.**

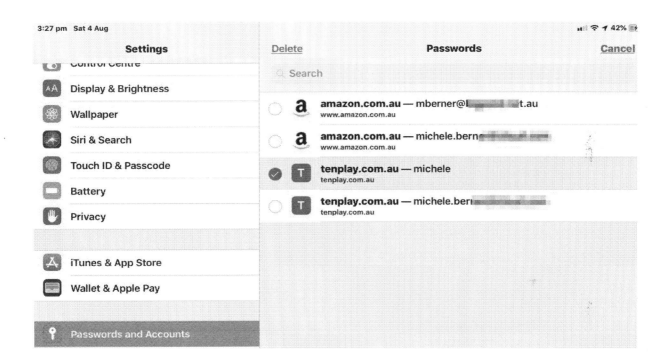

# Screen Time

This setting in the Settings app collects data on what you're doing on your iPad. It gives you an idea of how much time you're playing your favourite game or browsing through your Facebook feed, whatever your device. It can help you monitor your iPad usage and get control back of your life. For some people, using a device is very addictive, and you can basically decrease the productivity of your day.

Go to **Settings>Screen Time.**

**The Screen Time Dashboard**

This is where you get detailed information about how much time you're spending on all your devices, provided the devices all have the same Apple ID. The top pane displays the total time spent on your devices that day. The information is displayed as bar graphs which you can tap to get more information.

**Use Screentime passcode**: Enter a passcode–a different passcode to the one you use to wake up your iPad. If you have kids, and you want to control their screentime using the features described below, this passcode will be especially critical. You don't want your kids to bypass the Screentime and Downtime settings you've implemented. With a passcode established, this won't be possible.

**Share across devices:** If you enable this option, any Screentime and Downtime settings you set will be applied to all devices that are signed into the same iCloud account. It will then give you combined screentime usage reports across all the devices.

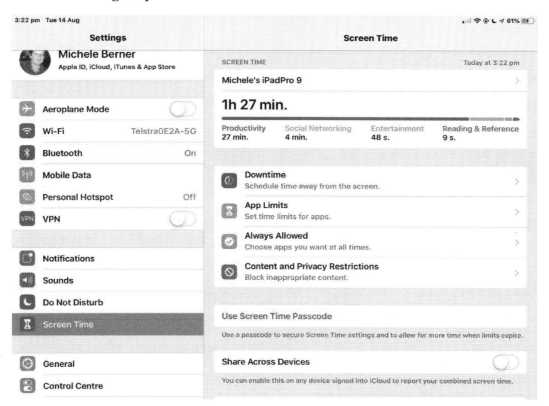

For example, the screenshot below tells me I have spent 37 minutes so far on my iPad today, as well as the time of the day I was using it. You can see a daily summary and also see the last 7 days of screen time with the total time used during the whole week.

Below that, I can see the **Most Used apps** from that day and the time spent accessing that app.

Under the App list, I can see how many times I've **picked up** my iPad, how often, and what times. This would be useful if you wanted to see where big chunks of your day had disappeared. You might then want to restrict access to certain apps at specific times of the day.

Below the Pickups section is the **Notification summary**. It displays how many notifications you've received, how often you receive a notification, and which apps are sending them. Tap an app which takes you to the Notification setting for that app. You can then turn off notifications for that app or change other settings like banner style or sound. Notifications are always distracting me – I need to stop what I'm doing and check it out, so this could be useful in turning off notifications for a specific app.

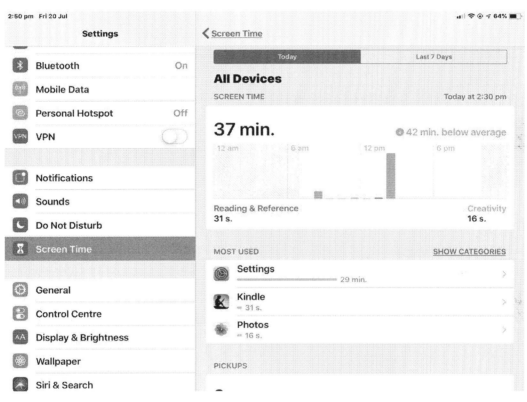

## Downtime

Downtime lets you schedule a time each day when your iPad is off limits. When you set Downtime, only the apps that you have approved can be accessed and everything else is off limits. This is meant to remove the urge to use your device and specific apps when you should be working, studying or getting distracted by notifications, which then make you open that app to follow up on the notification. You might use Downtime just before bedtime when you should be winding down before sleep and not checking your Facebook feed one last time.

**Note** that the Downtime settings will apply to any Apple device signed into iCloud.

1. To access downtime, create a passcode, enter it twice, and turn on Downtime. Once you set it, only approved apps will be available.

2. Set a start and end time, and each day, whatever you have blocked will be unavailable during those times – until you turn off Downtime.

3. Tap Screen time in the top left corner, and you're set. You'll get a notification five minutes before Downtime begins.

4. To set which apps are allowed to be used when Downtime is on, go to **Settings>Screen time>Always Allowed**.

When Downtime is active, all "banned" apps will be grayed out, and the only ones you can use are the Apps you set in the Always Allowed section. However, you can override this. When you launch an app that is grayed out, you can tap 'Ignore Limit.'

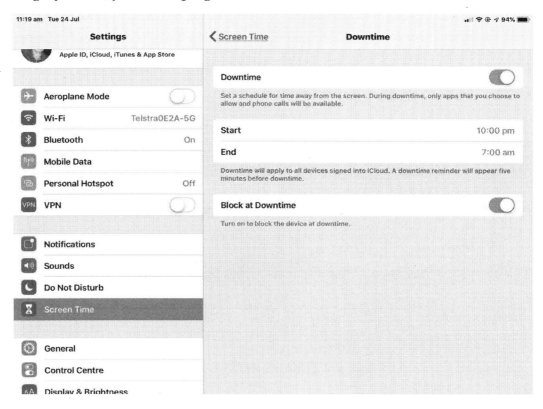

**App Limits**

You can place limits on entire categories of apps or just limit specific apps.

Once you are nearing your limit, you'll see a notification that there are "5 minutes remaining for the [name of the app] today." You can always go back into the App Limits settings and delete the time limit or reset it. Once you hit your limit, you'll see the **Time Limit Reached** screen instead of the app's content. You can tap "Ignore Limit For Today" or tap "Remind Me in 15 Minutes."

1. Tap **App Limits.**

2. Select, **Add Limit**. You'll see a list of all the categories you can restrict. Select the category you want, or select **All Apps and Categories**

3. **Once you have selected your category,** tap **Add** in the top right corner.

4. Use the scroll wheel to select how many hours/minutes you wish to apply to the category. If you have selected multiple categories, the time limit you set will apply to apps in all those categories. This becomes the total time you have allocated to use that category of apps. Once you hit the limit, you'll see the **Time Limit Reached** screen instead of the app's content.

5. To customise the days of the week to limit apps, tap Time; you can also adjust the time you set in the scroll wheel. Then **Customise Days**. You might want to allocate more time on the weekends, for example. Tap the day you want to change and adjust the time allowed. Tap the back arrow on the top left to accept the changes. Then, tap the back arrow again to get back to the App Limits screen.

6. To delete a limit, open the limit you have created and tap

**Delete Limit**.

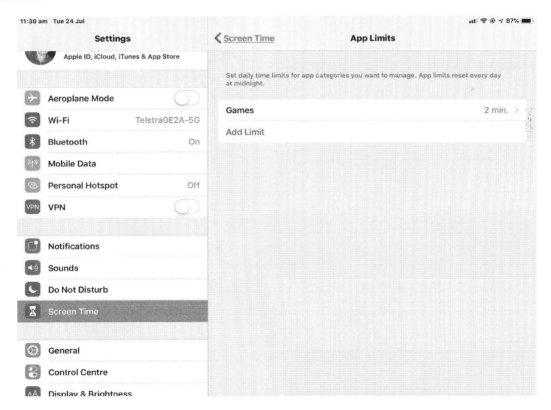

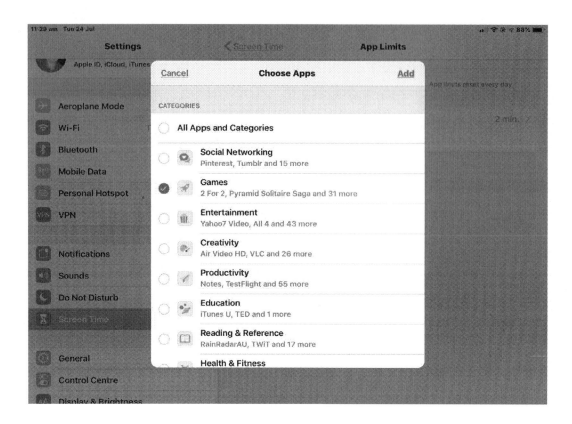

The image above shows that the whole Games category has had App Limits applied to it. If I want to play a game and remove the app limits, I can add that game to the Always Allowed section.

**Always Allowed**

In setting Downtime, you set a category or all categories of apps. I can turn off all the Productivity apps on my iPad. But what if I want to use a specific productivity app like Notes? You can turn on an app to always be allowed.

So, when Downtime turns off all my productivity apps at the time I set, I can still use the Notes app.

1. Tap **Always Allowed.**

2. From the **Choose Apps** section, tap the 'green plus' to add the app to the **Allowed Apps** section.

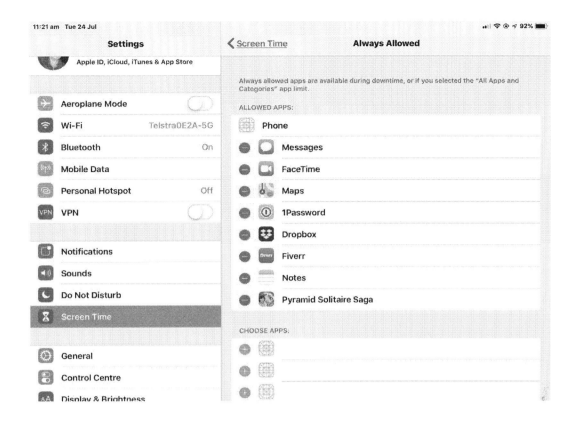

# Content & Privacy Restrictions

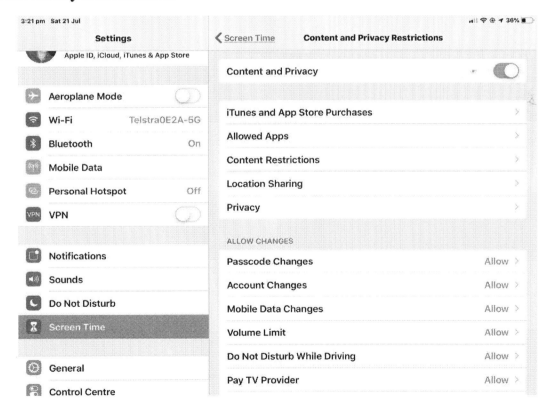

To make any changes to this section, enter your passcode, and then Turn on **Content and Privacy**. You can now make changes to the following:

**iTunes and App Store Purchases:** You can turn on/off the ability to install apps, delete apps, re-download apps and tune off in-app purchases. Perfect for kids to prevent them from making unnecessary and frequent in-app purchases or accessing any content that is not appropriate. You can also specify whether purchases and downloads require an Apple ID password.

**Allowed Apps:** You can turn on/off the apps that come pre-installed on the iPad. For example, if you want to turn off Siri and Dictation, FaceTime, Safari, Mail, Airdrop or the Camera, iTunes Store, Book Store, Podcasts and News, this is where you do it.

**Content Restrictions**: You can decide what content you'll allow for content: movies, TV shows, music, apps, books, news, podcasts. You can also set the ratings for your particular country. For websites, you can have unrestricted access. For **Limit Adult websites**, you can add sites that are always allowed and specify which sites should always be blocked. For **Allowed Websites Only**, you can grant access to all the child-friendly sites Apple has already provided (these can be deleted by swiping left) or add your own website as a whitelist.

**Location Sharing:** If you turn on this feature, you can share your location with family and friends in both the Find My Friends app and the Messages app. If you use the Home app, turn on this feature to enable location-based automation as well as a personal request for Siri when you use a Homepad speaker.

**Privacy:** This is where you can see which apps have requested access to built-in features like the microphone, photos, Bluetooth sharing, contacts, etc. When you tap a specific feature, you can see which apps have access to it and use the toggle to turn on/off access. You can also toggle Location services and determine which apps can have access to your location (Never, While Using, Always).

**Other Changes:** You can manage access to the following: passcode changes, account changes, mobile data changes, set the volume limit, and background app activities.

**Other features**

**You can set a passcode to use Screen Time**. A 4-digit passcode is useful if you have kids and they want to override any screen time settings. You can also use this passcode to allow more time when limits expire.

**Turn off Screen time.** If you don't believe this feature will be of any use to you, just turn it off. Any website, app or notification history will be deleted, and your digital behaviour will no longer be monitored.

**Combine Downtime with Bedtime mode in Do Not Disturb.** If you truly want no interruptions during the night, set both bedtime mode and Downtime.

**So, is Screen time worth the trouble?**

Screen time is a good way of monitoring your internet and app usage and gives you greater control over your digital health. After using it for a few weeks, I could see what apps I was using and overusing, so it gave me a greater sense of having my priorities all wrong or demonstrating my lack of productivity. It told me what I should be doing.

While **App Limits** appears very broad, you can use the **Always Allowed** section to bypass this to keep specific apps that you need to use all the time active when Downtime is enabled.

As a form of parental control, Screen time is worth using as it gives the parent a more informed idea of their child's screen use.

# How to recover or reset your Passcode

There are two passwords associated with an iPad. Your Apple account is the account you use when you are purchasing apps, movies, music, etc., on your iPad. Your apple ID is the first password.

The second password is the one your iPad may ask for when you "wake" it up. This password is commonly referred to as a "passcode" and usually contains four numbers. Recovering it will involve resetting your iPad to factory defaults – i.e. wiping it clean and restoring it from a backup, so make sure you have an iCloud backup first. (**Settings>Accounts > Passwords>iCloud>iCloud Backup**)

If you forget your password, the easiest way to recover it is to use iCloud to reset your iPad. The Find My iPad feature can reset your iPad remotely. You will normally use this if you want to ensure that nobody can open your iPad and access any personal information. You can easily wipe your iPad without using your iPad, which is an added benefit. Of course, you'll need to have to **Find My iPad** turned on for this to work. (**Settings>Accounts & Passwords>iCloud>Find My iPad** – Turn this **On**.

Go to https://www.icloud.com in a web browser.

Sign into iCloud when prompted.

Click on "**Find iPhone**".

When you see the map, click "**All Devices**" at the top and choose your iPad from the list.

A window with three buttons appears in the top-left corner of the map when the iPad is selected. **Play Sound, Lost Mode** (which locks the iPad down) and **Erased iPad**.

Just above these buttons is a list of your Apple devices that are currently registered. Verify that the device name is, in fact, your iPad. You don't want to erase your iPad by mistake!

Tap the **Erase iPad** button and follow the directions. You'll need to re-verify your choice and then your iPad will begin resetting. Note: To get this to work, charge and connect your iPad to the Internet, so make sure you plug it in while it is resetting.

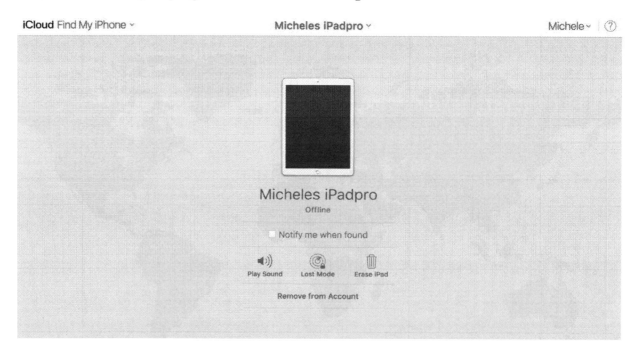

## Resetting your iPad to factory settings

If you've deleted problem apps, rebooted your iPad on multiple occasions, and are still having consistent problems, there is one drastic measure you can use. This will fix almost everything except actual hardware issues. Reset your iPad to 'factory default' settings. This essentially deletes everything from your iPad and returns it to the state it was in when it was still in the box.

The initial action is to backup your iPad. Choose **Accounts and Passwords** in the Settings app and then **iCloud.**

Select **iCloud Backup** from the iCloud settings and then tap the **iCloud Backup link**. This will back up all your data to iCloud. During the setup process, you can restore your iPad from this backup. If you were upgrading to a new iPad, this is the same process you would undertake.

Next, you can reset the iPad by choosing **Settings>General** and tapping **Reset** at the bottom of the General settings. There are several options for resetting the iPad. To set it back to factory default, tap "**Erase All Content and Settings.**" Before attempting the factory default option, you can try other settings to see if they resolve the issue.

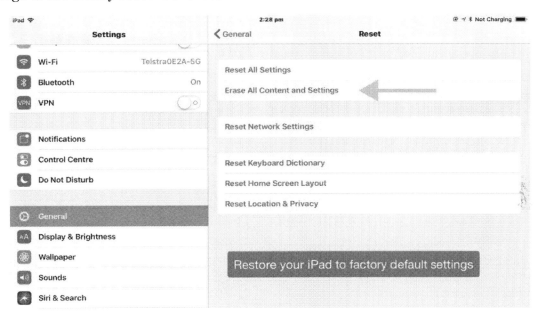

*Chapter 10:*

# Tips and tricks for your ipad

### Turn on the low power mode

Your iPad's battery life can be increased by using the Low Power Mode.

When enabled in the **Setting application> Battery** menu, the low power mode would reduce background activity, like downloading & receiving e-mails, to extend the battery life of your device.

### View media information in your Pictures

An essential info button has been added in the Photos application to view EXIF details of videos/pictures.

The info includes the size of the media, the device that took the picture, its location, etc.

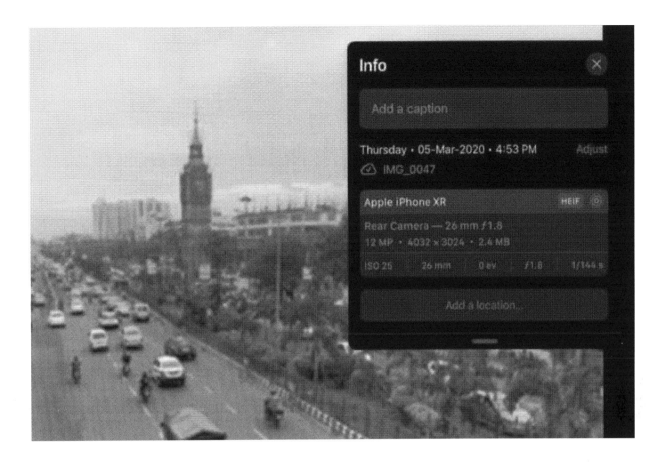

The iPadOS 15 not only allows you to view an image's EXIF details but also to change the date and time. You can now select pictures & videos and edit the date & time.

## Keyboard new shortcuts

Apple is offering a global keyboard shortcut for iPadOS 15, which runs across the whole OS.

| Split View | | Slide Over | |
|---|---|---|---|
| Enter Fullscreen | ⊕ F | Show Slide Over | ⊕ \ |
| Tile Window Left | ^ ⊕ ← | Move to Slide Over | ⌥ ⊕ \ |
| Tile Window Right | ^ ⊕ → | Move Slide Over Left | ⌥ ⊕ ← |
| Replace App | ^ ⊕ ↓ | Move Slide Over Right | ⌥ ⊕ → |
| Next Window | ⊕ ` | Next Slide Over Window | ⌘ ⊕ \ |

## System

| | | | |
|---|---|---|---|
| Go to Home Screen | ⊕ H | Siri | ⊕ S |
| Search | ⌘ space | Control Center | ⊕ C |
| Switch App | ⌘ ⇥ | | |
| Show Dock | ⊕ A | | |
| Show App Library | ⇧ ⊕ A | | |
| Quick Note | ⊕ Q | | |

**Write or draw on a picture**
- In the Photo application, click on a picture to see it on full screen.
- Click on **Edit**, then touch the Markup icon ⊚.
- Write & draw in the image with a variety of available drawing tools & colours. Touch the Add annotation icon ⊕ to magnify or add captions, text, shapes, or your signature.
- Click on Done to save it, or touch Cancel if you don't want to save the changes you've made.

# Chapter 11:

# Faq

Q: What distinguishes the iPad from the iPhone?

A: The iPad has a larger screen and a lower price. Another difference is that the iPad can be used for more than just iPhone apps. It can run its operating system, iOS. It also has some additional features with multitouch, a FaceTime camera, and more. A genuine reason to switch from an iPhone is if you need one for work or school.

Q: How do I charge my iPad?

A: The charging dock magnetically attaches so that your device will automatically be charged up to 80% every time you place it on the dock.

Q: Is an iPad too heavy to be portable?

A: The iPad weighs around 1.5 pounds, a little heavier than the iPhone, but it feels much lighter when you hold it in your hands.

Q: Will games slow my iPad down?

A: Games on the iPad need to take advantage of the extra processing power that the device has. The game cannot be too processor intensive and most titles will run like on any other device. If you notice a significant slowdown, your game is probably not optimized. Switching from an iPhone may help if your phone has a similar processor to a full-size iPad and there are few available yet for the iPhone or iPod Touch.

Q: How do I transfer music to the iPad?

A: You must have a short USB cable that connects your iPod to your computer. You can do this with iTunes. Also, you will need to connect the cable to your iPad or iPod before installing any software, like iSync, for syncing contacts and calendars. This process is not automatic; you need to do it yourself. If you are transferring music from an iPod without an SD card installed, then you will be more limited in how much music can be transferred at one time.

Q: How can I load music onto an iPad?

A: You can do this by first retrieving the music from your computer to your iPod, then transferring that to the iPad. If a USB cable is not attached to your iPad, you will want to turn off all security software on your computer and iPod. This way, both devices are allowed all permissions required for the transfer process, designed for both devices to be linked together. Also, remember that you cannot transfer these songs without iTunes installed on your computer.

Q: How do I copy MP3s onto my iPad?

A: When you purchase a song from iTunes or any other store program, it will automatically update your computer's DRM-free (.m4p) file. You can then transfer this to the iPad, playing it like a normal music player. Of course, you can also copy MP3s from a CD or an audio file.

Q: How do I access and use apps on my iPad?

A: You can move between apps by pressing the home button, swiping left or right, or double tap with two fingers to get to all of them at once. When you want to close an app, simply swipe it away using two fingers and tap the X at the top right of the screen. Swipe up on any app and tap "Clear" to delete one that's no longer needed.

Q: Do I need an App Store account to use an iPad?

A: Yes. You can create one via iTunes on your computer or by using the App Store app on your iPad. You will then be able to purchase and download apps right from your device. This way, you can avoid having to hook up your iPad to the computer whenever it needs a new app or more software updates.

Q: How do I add music, videos, and photos through iTunes?

A: When you use the iTunes program on your computer, you need to choose "Apps" from its menu bar so that there is an icon for them on the left of the screen. Then simply drag whatever media you want into this icon. Then connect your iPad (or iPod Touch) to your computer and sync it with your iTunes account. This way, you can be sure that all the media will be transferred from your computer onto the device.

Q: Can I add photos to a song or video?

A: Yes, if you have an iTunes account. Just drag photos into iTunes and then sync them to the iPad through the program, as explained above. When they are on there, you can scroll through your music and tap on a song or video while viewing its album cover art to add photos to it. Keep in mind that the photos need to be optimized for the iPad.

Q: How do I organize my apps on my iPad?

A: You can do this by tapping on your Home screen and choosing "Settings", then "General", then "Automatic App Updates" and turning off this function if you don't want it. Then choose "General", then "Siri", then "Use Siri" and turn it off if you are not going to use it. This will allow you to remove an app from your Home screen, but only when using Siri. You will still see all of your apps as long as they are signed in with iTunes or iCloud.

Q: Can I install apps to the iPad's SD card?

A: You can, but only if the apps are available outside the App Store. If they are only available there, you need to delete them and redownload them on a computer with iTunes. Apps will not be deleted from the iPad unless you format them or remove the app with iTunes and sync it again.

Q: What is AirPlay?

A: It allows you to stream media from your device to your television or computer screen. You need an Apple TV or a program like AirFoil (Windows) or AirVideo (Mac/Windows) installed on your computer or Apple TV 2 to make this work.

Q: Can I use the iPad with multiple accounts on the same device?

A: You can, but you will need more than one iTunes account to store all of your information.

Q: How do I delete photos from the iPad?

A: Here are a few different methods. You can remove all of your photos by going to Settings, then General, and scrolling down to "Erase All Content & Settings", then "Erase on Home Button". Or you can go to Settings, choose iTunes & App Stores and hit the album art icon at the bottom left corner of your screen.

# Conclusions

The iPad is a revolutionary device that has led to a fundamental shift in how individuals compute and interact with digital content. This iPad user guide explains why this device is such an incredible innovation.

With previous tablet devices, you were forced to poke and prod at screens as if they were made of Jell-O. On the other hand, the iPad is controlled using a simple interface that users can easily adjust through multi-touch gestures like pinching and panning gestures to zoom in or out on images or documents.

The iPad can also be used outside of your home or office. It is entirely usable in the sun with high-quality glare-free screens, and it features a brilliant yet thin black aluminium casing that looks just as good in a boardroom meeting as it does on your airplane.

Other than the ability to control content and applications by turning pages, swiping through digital magazines, or scrolling through digital news articles, you'll also be able to dig deeper into digital content thanks to a host of new multi-touch gestures featured on the iPad: pinching and spreading fingers change the zoom level; sliding fingers move forward or backward; and rotating fingers show the location of various application icons.

There are more than 300,000 applications already available for download from Apple's App Store, ranging from productivity-enhancing applications like Adobe Photoshop Touch and Pages to light-hearted games like Fruit Ninja and Poker Face. While some applications are free, the vast majority of these apps are not free but cost between $4.99 and $9.99. This means that there is an entire library of rich content you can explore for free on your iPad -- movies, TV shows, music, books, magazines and newspapers you can download from the Internet.

The iPad has an impressive display and high-resolution graphics. The display is 9.7 inches, which is large enough to view photos, documents and web pages with stunning clarity. And the resolution of the screen is 1024 by 768, which means that digital content on the iPad looks sharp even if it was originally displayed on a notebook or desktop computer.

The iPad's battery life lasts just over ten hours on a single charge, giving you plenty of time to read an entire book or watch an entire movie when you're out of the office.

The iPad charges using a power adapter plugs into an electrical outlet (like your TV). The adapter also acts as a stand for the iPad. The back of the charging adapter is reversible, so it can be used in over 150 countries.

The location of the Ethernet jack is on the bottom of the device, which means you can use your iPad with a wired network connection even when it's propped up in its stand on your desk or table. This can be crucial if you need to download large files quickly.

The Apple iPad connects to wireless networks using Wi-Fi 802.11n wireless networking (a.k.a., Wireless-N) and Bluetooth 2.1 + EDR wireless technology.

Made in the USA
Monee, IL
03 November 2022

17040207R00066